WORK ORGANIZATION AND ERGONOMICS

WORK ORGANIZATION
AND ERGONOMICS

Edited by

Vittorio DI MARTINO and Nigel CORLETT

David BUCHANAN
Peter CRESSEY
Juan Carlos HIBA
Felix SCHMID
John WILSON

INTERNATIONAL LABOUR OFFICE · GENEVA

Di Martino, Vittorio; Corlett, Nigel (eds.)
Work organization and ergonomics
Geneva, International Labour Office, 1998

/Manual/, /Work organization/, /Ergonomics/, Productivity/, /Enterprise restructuring/,
/Developed country/, /Developing country/. 13.03.2
ISBN 92-2-109518-5

ILO Cataloguing in Publication Data

Printed in Switzerland   WEI/SRO

Preface

As society and enterprises undergo major transformations, progress to better working conditions is often called into question. Many in business and in the community are asking questions, not necessarily new, but with renewed urgency, such as "what is an enterprise for?" Does it have a purpose beyond maximizing profits? Is there at least a satisfactory correspondence between how people work and what they consider the meaning and purpose of that work?

At the same time enterprises realize that it can no longer be "business as usual". Many are asking how they can reinvent their organization to achieve success in the face of continuing substantial changes in the nature and intensity of global competition. What organizational model should they follow or develop? What values should guide their home operations as well as their operations in diverse countries and cultures? How can enterprises ensure that people's perceptions of work keep pace with developments in the organization of work or, vice versa, how can enterprises ensure that the organization of work keeps pace with people's perceptions and expectations?

Examining the literature and looking around us we see that there is not a single model followed, no "one best way". Each company makes its own unique choice of purpose and values, and has its own model of critical business processes from which it derives its own measures of success. These choices are conditioned by many factors: the demand of stakeholders, the global market-place, rapid technological changes, community and public pressure, and government and international regulations.

We can, however, discern some common elements in what enterprises are doing to humanize work in their own workplaces and in the global market-place. Some of the approaches used are not new, but are becoming more urgent and pervasive precisely because of the social and economic challenges confronting enterprises. Others are new, as enterprises must continuously innovate and reinvent themselves to survive.

New experience from a growing number of leading organizations shows that it is not only possible, but necessary, to combine human resources, new technologies and quality environment in a positive-sum game to achieve higher competitiveness and success.

Based on such experience, this publication provides concrete tools to managers and workers to pursue this innovative approach. By presenting the most advanced techniques in the area of work organization and ergonomics in a simple and straightforward way, the book provides a much-needed, operative instrument both for developing and industrialized countries.

A multidisciplinary team of experts has been involved in a common project towards the production of this book. Although it should be seen as the result of the joint effort of the team, each of the experts was particularly responsible for certain issues or activities.

- Vittorio Di Martino, Chief of the Working Time, Work Organization and Technology Section, Conditions of Work and Welfare Facilities Branch, ILO, Geneva, designed and managed the project and was responsible for the general editing of the manual.

- Nigel Corlett, Emeritus Professor of Production Engineering at the University of Nottingham, United Kingdom, and scientific adviser to the University's Institute for Occupational Ergonomics, coedited the manual.

- David Buchanan, Professor of Organizational Behaviour at Leicester Business School, De Monfort University, United Kingdom, was primarily responsible for chapters on *Flexible work groups and multi-skilling* and *Strategic organizational issues*.

- Peter Cressey, Senior Lecturer in Social Sciences at the University of Bath, United Kingdom, covered the issue of *Workers' participation* throughout the various chapters of the manual.

- Juan Carlos Hiba, Senior Specialist in Ergonomics in the Conditions of Work and Welfare Facilities Branch, ILO, Geneva, was primarily responsible for the chapter on *Well-planned buildings and premises*.

- Felix Schmid, Senior Lecturer in Engineering at the University of Sheffield, United Kingdom, was primarily responsible for the chapter on *Layout of equipment and production flow*.

- John Wilson, Director of the Institute for Occupational Ergonomics at the University of Nottingham, United Kingdom, was primarily responsible for the chapters on *Task analysis and work design* and on *Workplace design*.

This book is for action. It is intended as a practical tool for all those wishing to change their enterprise into a better place to work and a more efficient organization. It is hoped that it will facilitate the search for immediate, simple-to-handle and low-cost solutions within enterprises wishing to become more modern, effective and socially aware.

F.J. Dy,
Chief,
Conditions of Work and Welfare Facilities Branch,
Working Conditions and Environment Department

Contents

Case-studies

Figures

Tables

Acknowledgements

We thank the publishers, individuals and organizations indicated in the text and in the references at the end of each chapter for their kind permission to reproduce tables, figures and case-studies.

Permission to reprint or quote from publications, subject to special mentioning, has been obtained from: Addison-Wesley Publishing Company Inc.; The British Psychological Society; Blackwell Publishers; Butterworth-Heinemann; Elsevier Science Ltd.; European Foundation for the Improvement of Living and Working Conditions; *European Industrial Relations Review*; European Trade Union Institute; *Financial Times*; Greenwood Publishing Group, Inc.; Her Majesty's Stationery Office; The Human Factors in Reliability Group; IFS International Ltd.; Ingersoll Engineers Ltd.; Routledge International; Thomson Publishing Services; Taylor and Francis; John Wiley & Sons, Inc.; and John Wiley & Sons, Ltd.

Special appreciation is due to David K. Rowley for the layout and graphic design, and to Kristine Falciola for the processing of the manuscript.

DESIGNING INNOVATIVE WORKPLACES   1

Vittorio Di Martino

Work organization and ergonomics

In the last 20 or 30 years, knowledge about how to run an efficient company has undergone profound change. To specialists the trend has been evident for even longer, but to many people, concerned with the day-to-day problems of business and industry, these developments come as fresh, new thinking, and have upset many traditions. This publication draws together some of the basic changes now being applied by enterprises around the world, and explains briefly what they are and how an enterprise may start to benefit from them.

The book applies particularly to a range of practical aspects of the operation of enterprises where knowledge from research and applications of ergonomics and work organization is exploited. Both these subjects deal, essentially, with matching the work environment to the needs and characteristics of people. They argue that if people come to work to manufacture things, for example, then the objective in designing their work – and the organization within which that work is done – should be to remove all obstacles which limit people's ability to get on with manufacturing.

This may seem obvious, but many traditions in work design and organization actually prevent a person achieving this objective. Ergonomics can demonstrate how often people have to reach excessively; pull, push or lift at the limits of their capacity; endure poor working climates; or do jobs so uninteresting that paying attention is a strain in itself.

Work organization shows a similar lack of modern knowledge. Payment schemes can restrict output, while organizations can create opposition between manager and worker which militate directly against high-quality output. Work in the last quarter of this century has instead demonstrated that it is possible to create human-oriented work organizations and ergonomically sound environments leading to major gains for both the workers and the company.

The integrated approach of work organization and ergonomics is a powerful means to increase productivity while improving the quality of work within enterprises. Improved work organization and ergonomics help to increase flexibility, production output and quality, meet customers' demands and adapt to technological innovation. At the same time, both contribute to reducing accidents and errors, undue stress and strain, absenteeism and turnover, and to spurring job satisfaction, morale and commitment of the management and workforce, often involved in delicate processes of transformation and change.

A new kind of enterprise is appearing, with competent and engaged workers who take greater responsibility for production, deliveries and performance. But such enterprises do not grow all by themselves. The right conditions must be created by enterprise management with a vision of new roles for their employees.

Greater efficiency and competitive strength can be attained by reorganizing the work and improving the working environment.

The approach

This publication is intended to show how to exploit the potential of ergonomics and work organization to achieve better working conditions and productivity by means of an integrated approach. It uses the complementary nature of the two disciplines to enhance this potential. Figure 8.1 at the end of this book illustrates the book's structure and how to use it to design for quality work.

When we speak of working conditions, this embraces the physical conditions of the workplace, the relationships between people and their equipment, and the organization in which they work. Organization includes the management structure and also the authority, the responsibility and the flexibility people have at their workplaces. It will be shown later that these are key features in the organizational environment to help people to work better, and be more productive.

CASE-STUDY 1A

ABB, Sweden

At the ABB plant in Västeras, Sweden, where electrical push button controls are manufactured, a development project was started in 1988. Its purpose was to increase efficiency and flexibility in the company by reorganizing the nature of the work in the assembly group. The new system was based on the following changes:

■ **A new work structure**

Instructions "from above" have disappeared, as have specialized functions. Instead, tasks that previously were outside the assembly group have been integrated.

■ **Responsibility**

New duties were successively imparted to the group. All members of the group were trained for their new tasks to enable job rotation within the unit. The group has overall responsibility for quality, operations, quantity and on-time delivery. The group is also responsible for inventory, capacity modification and competence.

■ **Work environment**

Equipment, workplace, flow, layout were all reviewed. Improvements were minor, but carried out systematically, based on the group's priorities.

■ **Individual development**

The individual was not given a specific job but was constantly trained for new duties. People climb the competence ladder to suit themselves, circulating among the various jobs within the group.

■ **Customer-oriented flow**

A new programme was introduced to reduce production cycle times and increase the service ratio. The employees were informed and trained for customer-oriented flow.

In less than three years (1988-1991), dramatic results were obtained. The company benefited from reductions in throughput time from 12 days to 1.5 days, in re-work from 15 per cent to 1 per cent; in absenteeism from 14 per cent to 4 per cent; in employee turnover from 39 per cent to 0 per cent; and a productivity increase of 14 per cent. Employees enjoyed a better working environment, greater job variety, better working relationships and increased wages thanks to higher productivity and levels of qualification. Altogether the company made annual savings of US$275,000.

Source: ABB Sweden, 1991.

Productivity is often presented as a simple ratio, such as output per worker hour. But this is too limited a definition. One company had 100 per cent labour turnover per year, due to poor working conditions. This is an enormous expense, but is not shown in the above productivity formula. Today, advanced enterprises will have a series of ratios for performance, monitoring not just output but re-work and scrap, turnover, machine down-time, work in progress and many others. Productivity to them is keeping such ratios within bounds, and continually improving them. Their target is to use their resources as efficiently as possible to achieve the objectives of the enterprise, and maximum output per worker hour may not be one of these.

Problems are interrelated – solutions are multiple

Problems in this area are often interrelated, as are solutions. Stress, for instance, is a growing cause of concern within companies. Absenteeism costs British business UK£5 billion a year. Anxiety, depression, neuroses, and alcohol and drug problems associated with stress are responsible for the great majority of cases of morbidity, disability and medical care use in the United States (see ILO, 1992, p. 13).

Since the factors provoking stress are multiple, action to combat stress also requires a variety of types of intervention. Particularly effective, in this respect, are measures to eliminate or reduce stress by improving work organization. People can be stressed from trying to do jobs over which they have poor control. Deadlines or output quantities may be unrealistic in relation to the available supply of parts or quality requirements. Poor tools and equipment are another source of difficulty, and added to these may be a poor standard of heating, ventilation, lighting or other aspects of the physical environment.

To combat such difficulties, they must all be dealt with by a combined attack, not just by specialists, but including the people who are doing the work itself. As the actual creators of the output and also the people who experience the problems personally, they are an important source of knowledge and ideas.

Combining actions of this type has proved very successful.

People, rather than machines, are the key factor

Both ergonomics and work organization deal, although from different perspectives, with the interface between individuals, or groups of individuals, within organizations and their occupations, the equipment, the workplace, the environment and systems. Workers are therefore the focus of attention in the two disciplines. This is consistent with the fact that, in the most advanced enterprises, attention is shifting from high technology matters to work organization issues and the role of people.

Against the traditional management philosophy and engineering principles which attempt to "design out" human skills on the road to full automation, it is becoming evident that enterprise strength does not come from the use of solutions based on advanced technology alone, but is based on the introduction of particular forms of work organization and the particular management of the complete production chain.

An effective response depends upon mobilizing all the skills and knowledge of the workforce, decentralized decision-making, more collaborative and participative forms of work organization, and computer systems which are designed for increased user control and greater transparency of the whole production process.

Even where the technological push is in full development, an apparent paradox is emerging that, as more technology is introduced, people become more critical to the running of increasingly sophisticated and vulnerable systems.

Strategic and immediate action are compatible

This publication is "action oriented". It is intended to promote and encourage direct action at the workplace, action to be undertaken as soon as possible. But action is often not easy to undertake because of lack of resources, resistance to change within the enterprise, conservatism or an unfavourable environment. The book proposes, therefore, a step-by-step approach based on progressive improvements, simple procedures and low-cost solutions. Sometimes the introduction of improvements is delayed, waiting for the "great" opportunity which never comes. Instead, the manual highlights the potential of a proactive approach through the continuous implementation of low-cost improvements using ergonomic techniques and work organization procedures that may actually pay for themselves very quickly in terms of increased production and better product quality.

These improvements, for example, can directly increase productivity by eliminating or combining tasks, by permitting more rapid task execution, by shortening handling distances and by reducing the likelihood of errors. Such improvements also reduce the fatigue of people and permit more rapid recovery, which indirectly contribute to productivity. Finally, many of these improvements affect people's motivation and thereby the likelihood of production increases and higher quality.

Proceeding by stages does not mean that action should be exclusively targeted to immediate benefits and short-term solutions. These benefits can be consolidated and magnified only within a long-term strategy of development and improvement of work organization and ergonomics. The most successful enterprises are those which are able to develop long-term action in both fields.

Both industrialized and developing countries can benefit

This publication is targeted to enterprises in both industrialized and developing countries. In industrialized countries, the importance of ergonomics and work organization is widely recognized, although within enterprises, sometimes advanced enterprises, the concrete application of the lessons from these disciplines is far from being as widespread as it could be. By looking at the interdependence among various environmental, organizational, technological and human factors within complex settings and sophisticated systems, this book may help to solve problems which can have major effects on the efficiency and productivity of companies in these countries.

This book may also prove particularly useful in countries moving from planned to market economies. Widespread privatization, the breakup of state monopolies, price liberalization and monetary and financial reforms often generate situations in which massive changes are necessary, in terms of technology, restructuring of enterprises, and the search for higher productivity and competitiveness. There is a lack of adequate expertise on how to handle these changes in an effective way at a local level, and a considerable danger that these changes worsen the conditions of work without achieving the expected improvements in efficiency and productivity. Ergonomics and work organization may provide the right response to these types of problems.

CASE-STUDY 1B

Paper Products Ltd., United Republic of Tanzania

Paper Products is an expanding factory due to the growing demand for its products, which include paper packaging materials and labels. The enterprise employs about 45 people.

Up to 1993, the factory building had no windows except a few louvre blocks in the top part of the walls, hence very little natural light entered the factory. Three hundred and seventy-six tube lights were installed to supply artificial light to the production lines. However, in spite of the artificial lighting, the production lines were quite insufficiently lit and the workers were often complaining of darkness. Whenever the electricity went out, which happens frequently, most of the production area became very dark, forcing production activities to stop.

In 1993, an intervention programme was carried out by the ILO in the United Republic of Tanzania within small and medium-sized enterprises aimed at improving working conditions and productivity (WISE methodology).

During the WISE course, the Paper Products management was advised to install translucent sheets in order to make use of natural light. The advice was accepted and within three months after the course, 16 translucent sheets, each 1 metre by 3 metres, were installed.

As a consequence of this intervention, the company makes a saving of US$72 each day in electricity bills. The company is also experiencing reduced expenditure on tube lights replacement. In addition, the management calculated that about 15 hours of effective working time which were wasted every working day due to insufficient lighting are now regained to production as a consequence of adequate lighting.

Finally interviews with workers at the factory showed that, in their view, working conditions and environment, safety, job satisfaction, quantity and quality of work, and working speed are "better" or "much better" than before.

Source: Lauwo, Kitunga and Tornberg, 1994.

Newly industrializing countries can also especially benefit from this publication. Their development is often very rapid and change occurs at a pace which sometimes makes it difficult to balance all factors of change properly. Technological progress may not be matched by adequate improvements in the way work is performed and in the conditions of work. However, these are not contradictory elements but parts of the same "package" to achieve efficiency and competitiveness. Work organization and ergonomics may be the linking factors between social and technological developments.

In developing countries, working conditions and ergonomics are also receiving growing attention. Experience in these countries has shown that applications of these two disciplines can prove very useful and successful when these practices are developed at a local level. In this respect, this book is an "open" instrument, a guide within these areas with no prescriptive intention.

The target enterprise

This publication is directed primarily at a small or medium-sized, medium-technology enterprise aiming at high performance and people development. Larger enterprises may certainly benefit from it, especially from a perspective of decentralization and creation of autonomous operating units.

It could be particularly useful to a company that is, or is trying, to become flexible, quality-oriented, participatory, customer-oriented, healthy, competence-based and committed.

A flexible enterprise

Flexibility has become a key factor of success in modern enterprises. It is the ability to adapt quickly to changing situations and demands that often determines whether an enterprise will survive or not in turbulent markets and in unpredictable environments. The response of the enterprise will therefore be targeted towards:

Product flexibility – the ability to introduce and produce novel products or services or to modify existing ones.

Mix flexibility – the ability to change the range of products or services.

Volume flexibility – the ability to change the level of output.

Delivery flexibility – the ability to change delivery dates.

In order to achieve these targets, the enterprise will enhance the flexibility of all its internal resources and, in particular:

Workforce flexibility – this is a workforce which is multi-skilled, multi-role and capable of combining different functions, thus allowing for higher degrees of mobility, integration, interchangeability and, therefore, productivity. A workforce which is flexible is one which requires a special effort in education and continuous training.

Technological flexibility – machines which can perform or be rapidly adapted to different functions, machines which are adapted to people, not vice versa. New technologies which allow for multiple options in the design of organizational systems.

Organizational flexibility – organizational mixes which can be easily remodelled according to varying needs and circumstances.

A quality-oriented enterprise

For this enterprise it is essential that its products and/or services meet the demands of those who actually use them. In a market which is increasingly global and competitive, with customers and users becoming increasingly sophisticated, quality is a key factor for competitiveness and survival. By improving quality the company can increase its income from sales, reduce the cost of scrap and re-work of detected faulty products, as well as the cost incurred by the undetected ones which reach the customers. This emphasis on quality requires a completely new organizational approach and the use of all the human resources within the enterprise. Quality thus becomes:

A strategic issue – quality needs continuous improvement and this can be achieved only if quality is considered a strategic issue, central to decision-making within the enterprise.

A management issue – as a strategic issue, quality becomes a management issue which necessitates top management support, as well as full commitment from all management levels.

A systemic issue – quality thus becomes an essential element of the company organization and culture, which requires preparation and extends to the entire production process and to all workers within the company.

CASE-STUDY 1C

The Hay Employee Attitude Study

The 1991-92 Hay Employee Attitude Study, based on survey work with more than 1,000 organizations in the United States; representing 2.5 million employees, found dramatic differences, on the issue of participation and commitment, between the surveyed companies and a group of 20 companies identified as "**leaders**" because of their performance and reputation.

In particular, employees in the leading companies were:
- prouder to work for their company;
- more involved and given greater authority;
- more respected and treated fairly;
- more informed about the company;
- more confident about the ability and credibility of top management;
- much more positive about quality.

In these leading firms, moreover, middle management – the group so disaffected in the survey as a whole – viewed their companies very favourably. They saw a deep and enduring commitment to quality, and they believed in top management's leadership ability.

Source: Work in America Institute, 1992.

A participatory issue – as a consequence of this internalization, quality demands open communications, greater worker involvement and the creation of high-trust relations within the company. As individual monitoring of quality is only possible with knowledge-based jobs, the worker must be a skilled, better a multi-skilled one, and receive extensive, continuous training.

A participatory enterprise

In the continuous process of adjustment and change, including technological change, that is important to any healthy organization and essential to the survival of a competitive enterprise, the involvement of all the parties concerned plays an increasingly important role. For both management and workforce, participation is seen more and more as a practical tool which helps the smooth running of the organization, the achievement of higher productivity levels, better working conditions and a better environment. It is perfectly functional for the trend in modern enterprises towards decentralized, flexible forms of organization and to adaptable problem-solving techniques.

A customer-oriented enterprise

Meeting customers' needs, first time, every time, is not only a slogan but a reality for a growing number of enterprises. In the successful enterprise, customers become, or are becoming, an integral part of the production process. Products and services are targeted to high quality, timely delivery, and adaptability and responsiveness to the requirements of each individual customer.

A new type of dealer emerges who is fully integrated, not only in the market, but in the company itself. Direct relationships are established between the company and the customers, their opinion and assessment of products and services actively sought, a continuous information flow maintained and a two-way communication system guaranteed.

CASE-STUDY 1D

Lucas Aerospace Participatory Approach

Since 1987, a new working system has been progressively introduced in Lucas Aerospace in the United Kingdom, based on the idea of "**changing a business through its people**". This system is targeted to improved product quality, improved customer services, and reduced delivery times. In this new system, workers participate in continuous improvement groups and play an active role in guaranteeing the quality of their product and in meeting the customers' requirements. Work has been reorganized across trades by enhancing flexibility and multi-skilling and by introducing self-inspection from operators. The system is highly participative with extended communication, consultation and involvement at all levels. Education, training and retraining are priority areas, with Lucas Aerospace almost doubling the expenditure of other companies in this area. In the period 1987-91, both the company and workers have greatly benefited from the new system in the following areas:

queues: no queues instead of a previous average of 2,000 batches queuing;

rectifications: the cost of rectifications has been reduced from UK£1800k to UK£339k;

customer returns: reduced from 1.6 per cent to 0.32 per cent;

scrap: reduced from UK£943k to UK£500k;

cost of quality: from 7 per cent to 4 per cent of sales;

quality of work: the work is much more autonomous and rich;

working time: increased flexibility, with workers largely responsible for the organization of their own working time;

remuneration: better financial conditions have been achieved;

staff status: has increased, workers have more to say in the running of the company;

training: a major area of intervention with higher costs being fully compensated by return benefits;

job security: the company, which has already achieved negligible levels of turnover, is now moving towards guaranteeing job security to its workers.

Source: Lucas Aerospace, 1992.

This approach requires:

* a flexible organization;
* a flexible workforce;
* high commitment from management and workforce;
* a participatory environment;
* a search for continuous improvements;
* new forms of linkage with the supplier and the customers.

A healthy enterprise

Do it in a clean, healthy way from the very beginning and everybody will benefit. The idea of operating clean, healthy enterprises is spreading all over the world. In many industrialized countries, and in some developing countries, there is nowadays a growing trend of improving the quality of people's working environments. Many enterprises are also understanding that it is not possible to manufacture high-quality products in unhealthy workplaces and environments. Moreover, there is an increasing preoccupation with the protection of external environments.

The healthy enterprise concept is not just the control of sources of pollution. The idea goes far beyond that.

Figure 1.1   A healthy company and its main elements

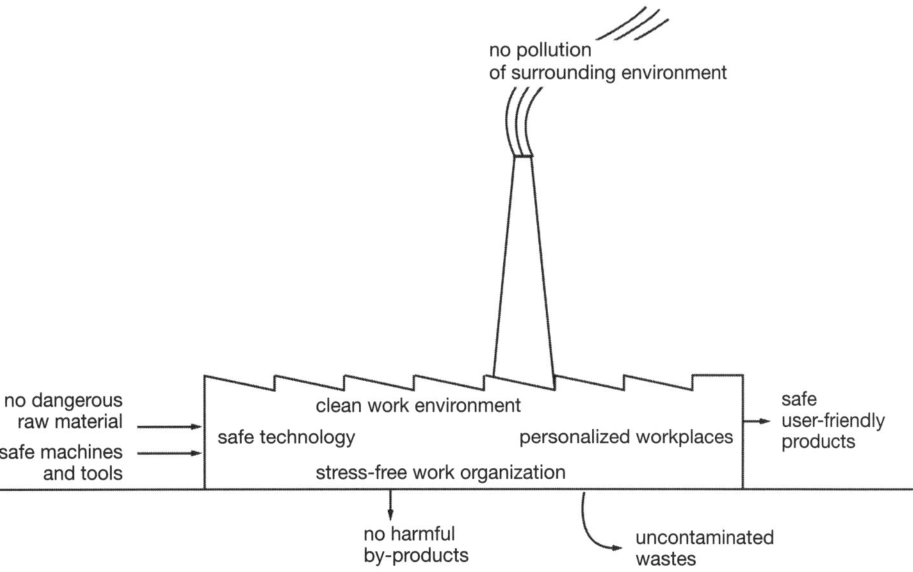

To start with, it means the use of no dangerous raw material and the purchasing of safe, productive machines and tools. It also conveys the idea of using safe technology that does not spoil the work environment. It means the possibility of eliminating harmful by-products and the generation of uncontaminated wastes.

Within the factory, a healthy enterprise will provide workplaces adapted to their workers, stress-free work organization, a clean physical work environment and the production of high-quality, safe products.

A competence-based and committed enterprise

Competence drives organizational effectiveness. The enterprise will, therefore, establish a personnel policy for promoting the continuing, long-term development of knowledge, skills and experience of workers. This policy will, in turn, lead to the introduction of multiple skills schemes where training is systematic, constant and integrated into the work itself.

Competence will also be promoted through the introduction of work in groups or teams, which often have responsibilities that go far beyond task execution, and by the implementation of flexible networks and cooperation, replacing bureaucratic hierarchies by open channels of communication and meetings to plan, coordinate and control work.

A competence-based organization will require the commitment of workers to productivity, quality and profitability, recognizing that these are prerequisites for income growth and employment. On the other hand, the view that long-term economic interests require dedicated workers, implies the commitment of the employer to their professional development and their job security, and the involvement of and cooperation between all parties in the organization.

Action at different levels

At enterprise level

Changes at this level affect the whole undertaking and may have an important impact on the workforce and in enterprise competitiveness. For example, consider the case of an enterprise discovering that the organization is not making an effective utilization of the company's human resources and that this is linked with problems related to working time. If, as a consequence, the company wants to introduce flexible working time, this is a major issue that will need careful planning, a full exchange of information between management and workers, the preparation of new regulations and instructions, probably a period of trial and readjustment, and then a full implementation of the approach.

At department level

The introduction of changes at the departmental or unit level can be the result of a strategic decision taken at a higher level. However, in a competence-based company, those changes can also well be the outcome of a decision taken at the department or unit level. Improvements can include the reorganization of working time arrangements, the implementation of training due to the introduction of new technology, the creation of autonomous groups, or the implementation of low-cost solutions on environmental and work organization issues. Decisions made at this level will certainly have an impact at both enterprise and workplace levels.

At workplace level

Changes at the workplace level can have a relative impact at a higher level, but nevertheless they are still important; sometimes these changes are crucial for the enterprise. Changes could be related to specific aspects of the production process, training of workers or other relevant matters. The implementation of changes at the workplace level provides an excellent opportunity for involving people. Changes at this level frequently have a direct impact on productivity and product quality. Depending on their nature and level of complexity, the implementation time may range from a few minutes to hours or even weeks. In an innovative firm, many changes can occur simultaneously. In a very innovative firm, the process is continuous.

The audience

The authors have assumed that the reader has the following characteristics:

*   Works for a medium-sized, medium-technology employer in industry or services and in a country with any of several levels of development.

*   A strong practical orientation; interested in what to do and how to do it.

*   Adequate, forward-looking and progressive attitudes; willing to consider the human, as well as production, implications of decisions.

*   Openness to a participatory approach to improvements.

*   Limited expertise in ergonomics and work organization.

*   Fairly severe financial constraints on action.

In addressing this audience, or rather this range of audiences, it has been found difficult to give specific, practical advice. This results from several factors, but the most important relate to the differing possible contexts in which the advice will be used (the specific conditions and situation at the workplace) and the many different potential backgrounds and perspectives of the audience (managers, supervisors, workers, engineers, etc.).

In order to overcome these difficulties, this book has been developed bearing in mind the specific needs of what we have called a ***design team***. An increasing number of enterprises address problems in teams.

A design team is a small group of people which has been assigned the task of considering, developing and introducing changes within a specific work context, including aspects such as improving ergonomics and work organization.

A design team is representative of the groups that will be affected by the changes it proposes. In particular, workers and supervisors must both be included.

The design team works in a participatory manner; it is given objectives, rather than instructions.

A design team has available to it the minimum necessary resources in terms of both time to devote to its task and access to any specialized expertise that is required. However, it is recognized that, in many cases, such resources are very likely to be modest. This manual therefore offers advice that is not necessarily costly, while being as self-contained and easy-to-apply as possible.

A design team has sanction at the appropriate levels of decision-making, and is able to make proposals with financial implications.

Design teams may be set up for a short time and with a single, limited purpose, or they may be part of a permanent, ongoing effort.

> **Design teams are crucial for innovation and change**

Because a design team has a representative composition and a participatory method of work, it can be expected to follow a strategy of adaptation of the advice included in this manual to local circumstances.

For similar reasons, a design team is capable of dealing with relatively complex situations and proposing more complete and sophisticated solutions. It has access to the full variety of skills of its members. The iterative nature of the discussion process favours consideration of a range of possibilities and adaptation of the action that is selected to the concerns of everyone involved. In this way, some of the dangers of one-sided or oversimplified solutions can be avoided.

References

ABB Sweden: "Operation Pushbutton", information material presented at the Workshop on Participatory Practices: What can we learn from the Swedish experience?, European Foundation for the Improvement of Living and Working Conditions, Dublin, May 1991.

ILO: *Conditions of Work Digest* on *"Preventing stress at work"*, Vol. 11, No. 2, 1992.

C.N. Lauwo, L.J. Kitunga and V.P. Tornberg: *African safety and health project*, unpublished report prepared for the ILO (Dar es Salaam, July 1994).

Lucas Aerospace: Presentation made at the 4th International Workshop on Employee Involvement: Innovative team working experiences in Europe, Brussels, March 1992.

Work in America Institute: *Work in America* (Elmsford, New York), Vol. 17, No. 1, January 1992.

TASK ANALYSIS AND WORK DESIGN   2

John Wilson

Designing effective and satisfying tasks and jobs

In the Chapter 1, we have looked in general at some aspects of the world of work, and discussed how the introduction of more reliable knowledge about people's abilities and needs can improve their contribution to a company, as well as improving that company's ability to satisfy its market and improve its commercial standing. It is now time to start demonstrating just how these advantages can be obtained.

This chapter discusses people's tasks and jobs in their immediate environment. It explains how physical effort is achieved and discusses its limits. Posture and the loads and strains introduced by bad postures are discussed, and some techniques to investigate these are given. Then we consider manual handling, still a very common activity and responsible for a large proportion of industrial injuries.

Mental work is introduced later in the chapter, to be followed by an introduction to job content, i.e. what should a job contain if it is to be done well.

Of course, no task or job can be studied and designed without referring to a wider context of physical workspace (see Chapter 3) or organization of work (see Chapters 4 and 5), but many improvements in work organization and work environment must start with ensuring appropriate task and job design. Figure 2.1 shows some of the key interactions relevant to work design.

Figure 2.1 Interactions in work redesign: Principles of ergonomics

Interactions for consideration when applying the principles of ergonomics to work design

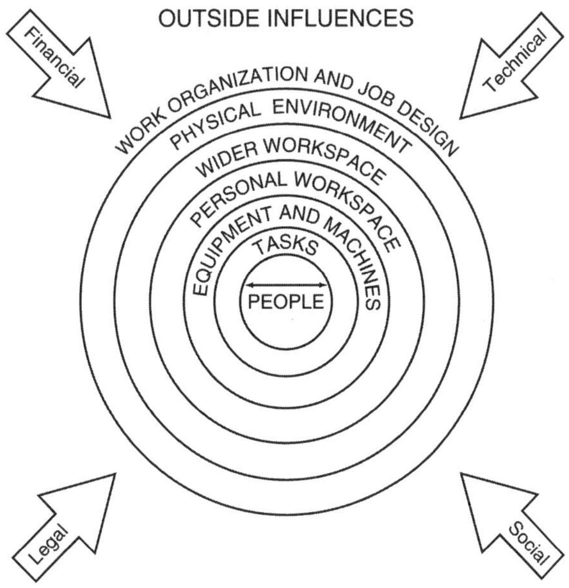

Source: Wilson and Corlett, 1995.

Definitions

Tasks are usually seen as the smallest useful description of a work activity, and can be defined as a goal to be achieved under certain conditions and with certain resources. Tasks are often expressed as "switch machine on", "change drill", or "type this letter". More detailed descriptions will account for conditions placed on performance and resources available, e.g. "insert piston in cylinder with minimum lubricant necessary before capping end, with mean cycle time of ten seconds" or "use A4 (letter) size paper without heading because it is only a draft".

Jobs generally consist of a number of tasks. We will see later that the more varied and numerous the tasks, usually the better the job. However, jobs contain more than just tasks, there are also degrees of responsibility and freedom, for instance. This level, which includes how much autonomy and decision-making power people have in their work, is their role.

Objectives in task and job design

The goals of task and job design are twofold: we want to produce tasks which are more feasible, effective and less harmful; and we also want jobs which are satisfying, productive and complementary to the rest of an organization's activities.

This is not easy to achieve, because individuals differ tremendously in their wants, needs and abilities and no two organizations are the same. Mismatches between individuals' abilities and task demands are therefore frequent. In designing new tasks and jobs, every effort should be made to avoid such mismatches.

Task description and analysis

The first stage in task or job design is a task description and analysis. This enables the design team to see clearly what must be done by the operator, under what circumstances, to what criteria and with what consequences.

Task description and analysis involves observing work, recording and describing it, and then analysing it. Since this analysis relates to the activities of people in the system at a given moment in time, as the design of the system progresses (or is upgraded after it becomes operational), then the analysis should be modified or upgraded accordingly.

Usually a task description and analysis for an existing job comprises the following steps:

1.   Observe and record what is happening at present.

2.   Define what functions are necessary to achieve the goals of the system or organization.

3.   Describe and represent what must be done at a task level in order to successfully complete those functions. This is the *task description*. Such a description is not of how the task is done now, but of what must be done, whatever the means employed.

4. Define tasks in terms of any critical characteristics, such as sequence, time for completion or time constraints, criticality, complexity, and their interrelationships.

5. Identify any equipment, interface or environmental constraints on safe and successful task performance; examine workspace, physical environment, social and work organization.

6. Identify key characteristics of the actual or potential employees or users.

7. Analyse the requirements of, and consequences for, the personnel carrying out the tasks in the particular context considered. This is the *task analysis* and it will include assessment of mental and physical overload or underload for the operators.

8. The task analysis will become the basis for decisions about allocation and division of functions, between people and equipment (e.g. the level of automation), for training, design of interfaces, equipment, jobs and systems.

In situations where a new system is being designed, there may be sparse or even no information available from existing task situations. In such cases, a **task synthesis** approach is taken, where the investigator must hypothesize tasks and the likely constraints on task performance. Information may be gathered from systems documentation and the design team as well as by extrapolating from existing comparable tasks. Several attempts are usually required before the final task structure is produced.

Task description

In order to be useful for subsequent analysis, a task description must be:

- complete and amenable to validation

- readable and simple to understand by non-trained staff

- clear and concise

- a communication about the task domain of common understanding.

To ensure that this is achieved, it is particularly important which features of tasks are recorded in the task description, prior to task analysis. The following are some of the key features to be considered.

Identity of component subtasks. A listing of the subactivities involved in a task.

Grouping of component subtasks. An organized, possibly hierarchical, listing of the sub-activities involved in the task, showing how subtasks cluster according to functional or temporal considerations.

Commonalities and interrelationships between subtasks. An indication of the extent to which component subtasks are employed, and are different aspects of the overall task, or of the way that subtasks cluster in terms of the information they require or the methods that are used in their execution.

Importance or priorities of component subtasks.

Frequency of component subtasks.

Success/failure rates of component subtasks.

Sequencing of component subtasks.

Time for completion of component subtasks.

Decisions made in the execution of component subtasks.

"Trigger" conditions for task execution. Many tasks begin execution simply as a result of the completion of a previous task or following a decision, though it is possible for the execution of a task to depend upon a particular stimulus or command originating within the task environment.

Objectives or goals of each task.

Information required by each task.

Information generated by each task.

Knowledge employed in making decisions.

Knowledge of system employed in performing task.

Task analysis

While there are many methods of representing tasks, and thus performing a task analysis, there are a few which are most applicable in field situations generally and which are widely usable. Table 2.1 illustrates an example by indicating in the left column the series of tasks to be performed by a port of entry operator and, in the right column, the potential impact of such tasks on the operator.

Table 2.1   Initial task analysis

Task analysis of port of entry operator	
Task	*Potential impact on operator*
1. Move pallet to side of conveyor.	1.1. Pushing or pulling heavy vehicle.
2. Enter shipping information from invoice.	2.1. Invoice and computer dialogue should be in same format. Dialogue should be written to be "user friendly" to materials handling operator.
3. Computer checks shipping information.	3.1. What happens if discrepancies are found – e.g. wrong codes on invoice, unordered parts, etc.?
4. Operator places carton on conveyor.	4.1. Height, reach, weight of carton. 4.2. Operator has to turn carton right way up to read information and ensure carton enters system correctly oriented.
5. Operator enters code from carton into computer.	5.1. Are the product code and a meaningful name on the carton in a position and size to make them readable? 5.2. Can operator remember 7-digit code long enough to enter it? 5.3. How does operator handle obviously damaged or wrong cartons?
6. Computer prints bar code for carton.	6.1. Is printer convenient to operator? 6.2. Does meaningful name appear on bar code sticker?
7. Operator puts bar code sticker on carton and checks code against box code.	7.1. Height, location and orientation of bar code are acceptable within wide limits – no problems. 7.2. Can operator compare two 7-digit numbers reliably? Meaningful names will help.
8. Operator places carton on conveyor.	8.1. Is release button convenient or does next computer input release carton?
9. Go to next carton (task 4).	9.1. Does operator have to put all cartons of one product code on to conveyor together? If so, he or she must hunt through pallet each time. 9.2. Does operator have to keep count of boxes? 9.3. If the last box has been coded does computer keep all of order "in bond" until it checks boxes against invoice for discrepancies? If so how are mistakes rectified?
Source: Adapted from Drury, 1983.	

Table 2.2 Processes used by people when doing jobs, and some points on how performance may be improved

Process	Activities	Specific behaviours	Some directions for improvement
Perceptual processes	1. Searching for and receiving information	Detects	Increase contrast between signal and background
		Inspects	Provide clear standards
		Observes	Assess size, lighting, contrast and position
		Reads	Use legible script and symbols; unambiguous
		Receives	Ensure it is a format to suit receiver
		Scans	Ensure that error value breaks pattern of normal values
	2. Identifying objects, actions, events	Discriminates	Maximize contrast between target and background
		Identifies	Use common coding or linking for associated features
		Locates	Use consistent arrangements between different parts of system
Mediational processes	1. Information processing	Categorizes	Clear criteria; enough time; job aids; formal structure
		Calculates	Job aids; appropriate displays
		Codes	Standardized scales; defined categories; job aids
		Interpolates	
		Itemizes	
		Tabulates	Job aids
		Translates	Consistency in coding
	2. Problem-solving and decision-making	Analyses	Training
		Calculates	Time; job aids
		Chooses	Job aids; consequences of choice
		Compares	Transfer mental to physical comparison
		Estimates	Maintain consistency by refreshing experience
		Plans	Reduce or sequence available choices; keep information on process state correct

Process	Activities	Specific behaviours	Some directions for improvement
Communication process		Advises	Face-to-face, on site
		Answers	Clear definition of questions
		Communicates	Simple language and sentences
		Directs	Objective requirements
		Indicates	Provide context and relevance
		Informs	Background, to aid priority and judgements
		Instructs	Training; follow-up/reinforcement
		Requests	Clear requirements, timing
		Transmits	Common codes; acknowledge receipt
Motor processes	1. Simple and discrete	Activates	Consistent differentiation between controls for functions
		Closes	Tactual/audible evidence of completion
		Connects	Positive evidence of completion; incompatibility of forbidden links
		Disconnects	Ready check if channel active; clear identification of different channels
		Moves	Job aids; control limits on movement
		Presses	Loading; indication of success of action
		Sets	Clear indication of targets
	2. Complex and continuous	Adjusts	Feedback, control characteristics appropriate
		Aligns	Indicator and target matched; control gain appropriate
		Regulates	Clear criteria; links between controlled factors overt
		Synchronizes	Control gain appropriate; amplify display of responses
		Tracks	Maximize acceptable target range; increase visible distance ahead of tracking point

Source: Based on Berliner et al., 1965, and Meister, 1985.

Design of tasks

Having set out what is needed to achieve the objectives, it is now necessary to design tasks which will enable people to undertake them.

It is very difficult to provide a detailed guide for designing work tasks; work goals, systems and equipment differ so much and any task design process must be relevant to the particular situation. However, the following basic principles should be taken into consideration when designing tasks:

- Tasks should be designed in a way that can be accomplished, that is, they should be conceived according to the capabilities and limitations of human beings.

- Tasks should be designed to be carried out with the least possible effort.

- Tasks should be arranged or combined in a logical way.

- It is good practice for the identification of human efforts and task sequence to involve experienced workers.

When designing tasks, it is important to identify specifically what the person has to do, so that the demands of the tasks are fully recognized. Table 2.2 will help in this by providing a summary classification of possible *task behaviours* in respect of the activities people are often involved in when working.

The proposals for improvement in the right-hand column of this table are not sufficient, nor exclusive to their categories. For example, in any mental activity more time is usually beneficial to success. Also, the availability of job aids, from pencil and paper, via instructor manuals or special tools to expert systems or simulation procedures or computers, can be applied widely. Training, too, is necessary, with refresher courses at intervals. Another common requirement is consistency. The use of standard codes in spoken or written communications, and standardized controls and displays across systems, can speed up responses and reduce errors in understanding and action.

Workload

One fundamental issue is the workload placed upon individuals by the tasks they carry out. When work is largely a physical activity, then the major consideration is to ensure that the physical demands of work are not greater than the capacity of the worker in those particular circumstances. With mental work – work involving attention or decision-making for instance – the need is to prevent both overload and also underload of the worker.

Physical and mental work will be considered separately now.

Physical work

The human body uses up energy in two basic ways. There is the use of energy to maintain postures that allow work to be done and to resist forces from gravity or from work equipment; this is *static work*. Bodies also use energy in the exertion of forces while moving; this is *dynamic work*. Figure 2.2 provides a simple illustration.

Static work

Problems from static work will arise if working postures are inappropriate, or are held for too long, or when the activities performed in that posture increase the stresses on body parts. Although not appearing to be as "hard" as dynamic work, it is important to realize that maintenance of any posture for long periods of time will lead to ineffective work, aches and pains for the workers towards the end of their shift, and long-term health problems.

The sites of muscular pain associated with static work can help to identify a poor match between work demands and the worker's capacity. Table 2.3 shows how this identification can be done.

Figure 2.2   Dynamic and static work muscular effort

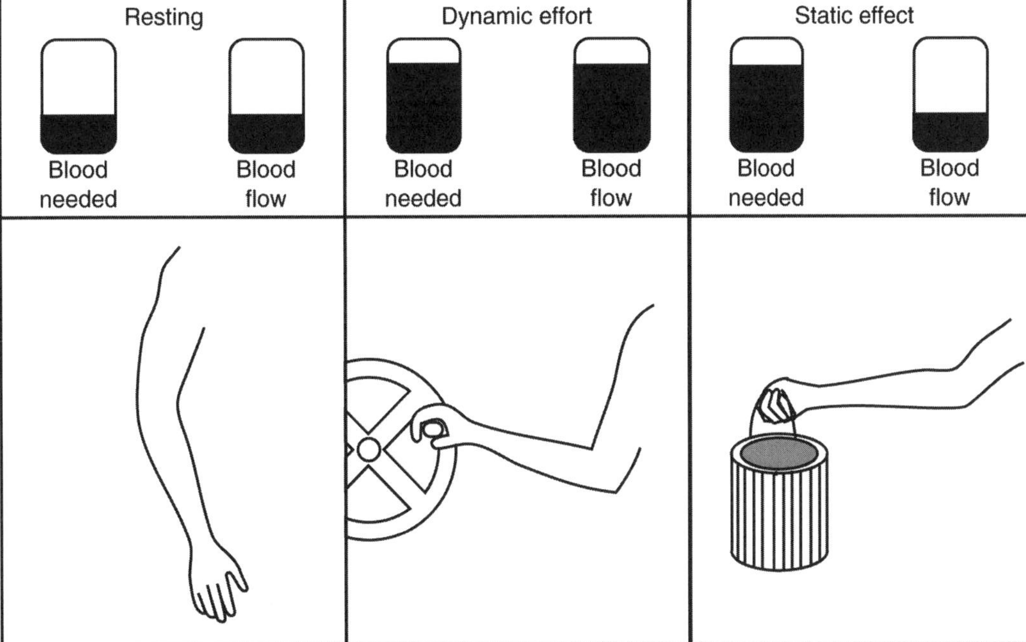

In static work, the muscles are exerting a steady force, remain tightly in tension and inhibit the blood flow needed to refresh them and keep them working. Thus, in a very short time, muscular pain is felt and the muscles must be relaxed. In dynamic work, on the other hand, the muscles alternatively tighten and relax, giving a chance for fresh blood to refresh the muscles and, indeed, helping this flow by the "pumping" effect of alternate pressure and relaxation. So dynamic work can be continued longer than static work.

Source: Grandjean, 1982.

Table 2.3 Operator postures that will lead to health (and efficiency) problems

Bad postures	Probable site of pain or other symptoms	Low-cost solution
Standing (and particularly a pigeon-footed stance)	Feet, lumbar region	Provide a stool
Sitting without lumbar support	Lumbar region	Provide a seat with lumbar support
Sitting without support for the back	Back muscles	Provide a seat with backrest
Sitting without good footrests of the correct height	Knees, legs and lumbar region	Provide adjustable footrest
Sitting with elbows rested on a working surface that is too high	Shoulder and neck muscles	Lower table height; increase seat height; provide adjustable seat
Upper arm hanging unsupported out of vertical	Shoulders, upper arms	Provide armrests
Arms reaching upwards	Shoulders, upper arms	Provide adjustable platform, lower point of work
Head bent back	Neck region	Provide seat with casters and reclinable backrest, move display to a lower position
Trunk bent forward; stooping position	Lumbar region, back muscles	Increase working height
Lifting heavy weights with back bent forward	Lumbar region back muscles	Use mechanical aids; teach safe lifting method
Any cramped position	The muscles involved	Redesign workplace
Maintenance of any joint in its extreme position	The joint involved	Bring controls closer to operator

Source: Adapted from Van Wely, 1970.

Body part discomfort mapping

Information about people's discomfort and the effects of poor posture is better obtained from a simple survey technique than from ad hoc questions. The investigation of body part discomfort by mapping the results from questioning people onto a diagram speeds up and improves its accuracy.

There are two approaches to mapping the development of discomfort:

(1)   At intervals throughout the working day people are asked to point to the site(s) of current discomfort on a body map (figure 2.3). They are then asked to rate the intensity of discomfort at each identified site on a 5- or 7-point scale, for example by marking a scale which is "anchored" at the 0 and 5(7) points by "no discomfort" and "extreme discomfort" respectively. These scores are plotted against time of day for each body site, dividing the scale into half points to give a 10- (or 14-) point scale.

(2)   At intervals throughout the day, people are asked to point out on the body map site(s) which are most uncomfortable, then those next most uncomfortable, and so on until no more sites are reported. This can be done by using an individual body map, but without the printed numbers, and asking each person to put a number in each marked off area according to the severity of the discomfort. By doing this, the person thus identifies a sequence of just-noticeable differences in discomfort, so it is unlikely that more than five or six will be recognized.

As they are identified they are numbered: 1 for the worst, 2 for the next worst, etc. By counting back from the not-identified (no discomfort) sites, counting these as zero, a scale of relative discomfort is developed.

Numbering, and the division for the body map, can be arranged to suit the problem being investigated. Prior discussion with those affected will allow the investigator to decide whether, for example, to mark off a section for the wrists, or whether the differing activities of each leg require separate sets of numbers for each.

This procedure is quick and gives results which are quite as effective as more complex methods.

Figure 2.3 Body part discomfort diagram

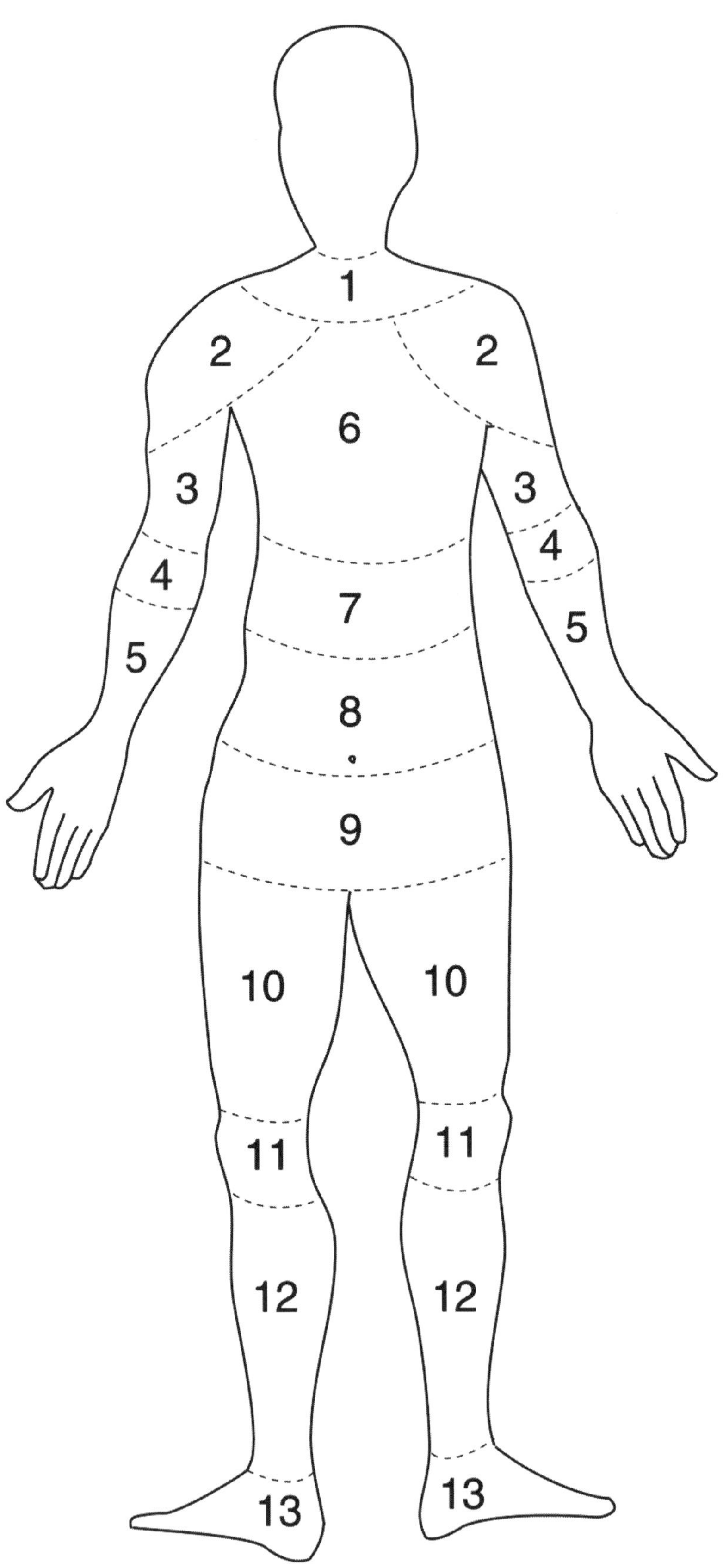

Source: Corlett, 1989.

Dynamic work

Although body movement is necessary to help prevent static work problems, excessive movements – especially when handling heavy loads – can also lead to health and performance problems. Heavy physical work still exists in many industries despite increased automation and the shift in work from blue-collar to white-collar jobs; people still lift and carry, push and pull loads.

Dynamic work which may cause problems, for the worker and for the organization, can be divided broadly into two sorts: (i) jobs where there is considerable expenditure of energy, and (ii) jobs which involve lifting and handling of loads. Particular risk factors over and above the task itself are workers' health, age, design of equipment, thermal conditions, length of work periods, and frequency of rest breaks.

It must also be borne in mind when using the results, recalling that the relationship between people and work is not the same as between a machine and its output, that there are factors other than physical effort which affect health and performance.

(i)   *Energy expenditure*

Two simple straightforward methods can help to assess the physical effort of a job.

The first is a method introduced by Brouha (1960), to use the "recovery pulse", the change in heart rate as a person recovers after some physical work. The worker sits down immediately after stopping work, and the pulses (P) counted for the last 30 seconds of each of the first three minutes of rest. Double each of these values (to estimate the pulse beats for each minute) to give P1, P2 and P3 and find the average (Pav. 1,2,3).

- If P1 to P3 > 10, or if P1, P2 and P3 are all below 90, then recovery is normal and the physiological workload is low.

- If the *average* of P1 over a number of recordings is 110, and P1 to P3 > 10, then the workload is not excessive.

- If P1 to P3 < 10 and if P3 > 90, then recovery is inadequate for the task and the person is overloaded.

A subjective method can also be useful, devised by Borg (1985). He asked people to indicate, on the scale given in box 2.1, their perception of their physical exertion (RPE). These perceptions were closely related to heart rates measured when they were on an exercise bicycle. The numbers, when multiplied by 10, approximate to the heart rates, but note that these are *not* the same as used in Brouha's method. There, the heart rates were approximations to the recovery rates *after* work, and so are lower than the peak rates during work. Also, the "maximal exertion" at 200 beats per minute of Borg's scale would be suitable only for young people.

Box 2.1   Rating of perceived exertion (Borg's scale)

6	**No exertion at all**
7	**Extremely light**
8	
9	**Very light**
10	
11	**Light**
12	
13	**Somewhat hard**
14	
15	**Hard (heavy)**
16	
17	**Very hard**
18	
19	**Extremely hard**
20	**Maximal exertion**

Both these techniques will give an estimate of the whole physical effort, which will include the effects of heat, since if the body is too warm, one of the mechanisms for reducing its temperature requires the heart rate to rise.

Readers should refer to the original publication for more details of these methods.

(ii)   Lifting and handling

Wherever possible, manual lifting and handling of loads should be eliminated. When this is not possible, it should be kept in mind that, in lifting and handling loads, it is not only the weight that is critical. The posture is an important variant as is the frequency of lifting and handling. The same load lifted with the trunk upright and the load hugged to the body can be within a person's capacity, but if lifted every few minutes or lifted with the trunk twisted or bent forward, it can lead to serious injury.

Some guidelines for manual handling are given in box 2.2.

Box 2.2   Ten guidelines on manual handling

1. **Avoid having to lift or transfer wherever possible. Make maximal use of aids (transfer boards, etc.), equipment (hoists, etc.) and gravity. Ensure you are familiar with how to use the equipment safely, and that it is in good working order. *Report* faulty or poorly designed equipment.**

2. **Do *not* attempt to lift or transfer a weight which *may* be beyond your capacity or is likely to be unpredictable without suitable assistance (mechanical or a colleague). If in doubt, put the object/person in a safe position and get assistance.**

3. **The number of trunk flexion movements required to perform a task and the range of rotation and side bending should be kept to a minimum.**

4. **Any requirements for a load to be supported by force applied over a substantial part of the work cycle should be minimized. This is even more important when the load or force being exerted is not close to the body, or the body is not upright.**

5. **Handling loads at heights below the knees or above the shoulder should be eliminated wherever possible.**

6. **Avoid using grasps which are not two-handed with the load or force evenly distributed between both hands/arms. Adopt grasps with maximal body contact area and where the upper arms and elbows are by the side, elbows are in mid-range and wrists are not twisted.**

7. **Clear away congestion and open up confined workspaces to ensure adequate space can be used to perform a task.**

8. **Where possible, select a colleague of similar height when performing two-person lifts or transfers.**

9. **Avoid jolting or sudden movements especially with load or force, or with trunk flexion or rotation.**

10. **Wherever possible use the trunk and upper limb muscles to stabilize the load or posture and use the leg muscles (especially the thigh muscles) to provide the force or movement required.**

Source: These guidelines were produced by E. N. Corlett.

Work postures

The critical importance of posture in relation to materials handling was referred to earlier. A procedure to assess the quality of postures, particularly when applying forces, is the OVAKO Work Analysis System, OWAS (1992), developed in conjunction with the Ovako steelworks in the mid-1970s. It is easy to learn and can be used by a wide range of the workforce to highlight aspects of work which may be hazardous.

Postures are observed and scored as shown in figure 2.4. An analysis form is used to assess each posture for acceptability or else remedial action. OWAS aims to identify postures where force exertions can be dangerous. Pushing, pulling or moving loads when the body is twisted or asymmetrically loaded are recommended for change.

The procedure is to observe the workers under study at intervals of 30 or 60 seconds, and record the postures and forces, over a representative period, say for 45 minutes. The scores are linked to the relevant phases in the work cycle by the final number.

From these data, the postures can be compared to the table of action categories of figure 2.5. Where force is involved, it is suggested that, as a guide, one less than the force score is added to the action category code of figure 2.5 to specify the action to be taken. Thus, if the posture action category is 2, as in the figure, and the force code is 2, then 1 is added to the action category, putting it up to 3, to be done as soon as possible.

The other useful feature from OWAS is the opportunity to identify the contribution of time spent in the different postures. If over 100 samples have been taken through the day, the numbers of times a harmful score for the back, arms or legs occurs can be added and converted to a percentage. Table 2.4 provides categories to recommend actions for various percentages of the different posture scores.

Figure 2.4   OWAS method: Classification of work postures

Items of the OWAS method and an example code for wiping during cleaning work

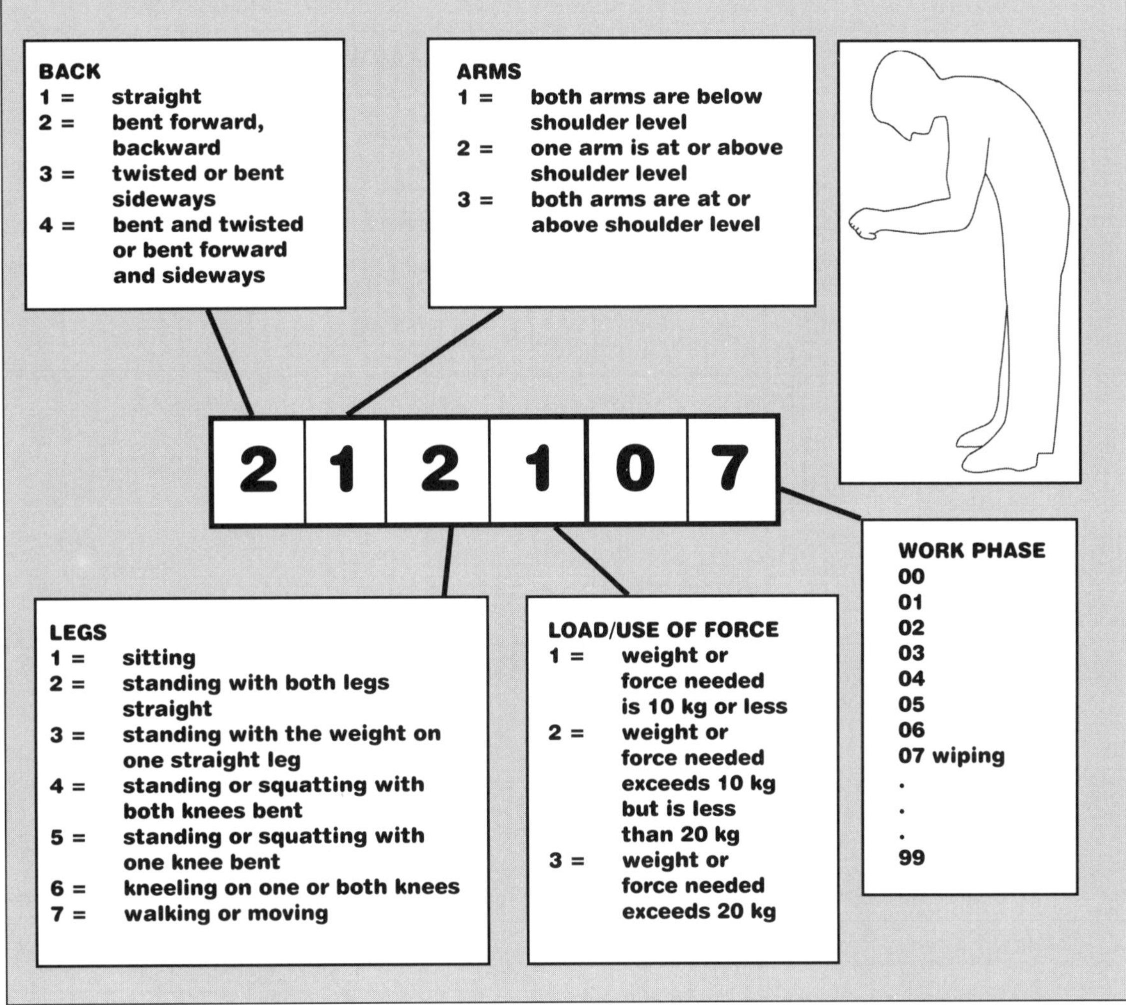

Source: Louhevaara and Suurnäkki, 1992.

Figure 2.5   OWAS method: Action categories for work posture combinations

LEGS postures are the column groups numbered 1–7; USE OF FORCE is the 1 / 2 / 3 sub-columns within each group.

BACK	ARMS	1			2			3			4			5			6			7		
		1	2	3	1	2	3	1	2	3	1	2	3	1	2	3	1	2	3	1	2	3
1	1	1	1	1	1	1	1	1	1	1	2	2	2	2	2	2	1	1	1	1	1	1
	2	1	1	1	1	1	1	1	1	1	2	2	2	2	2	2	1	1	1	1	1	1
	3	1	1	1	1	1	1	1	1	1	2	2	3	2	2	3	1	1	1	1	1	2
2	1	2	2	3	2	2	3	2	2	3	3	3	3	3	3	3	2	2	2	2	3	3
	2	2	2	3	2	2	3	2	3	3	3	4	4	3	4	4	3	3	4	2	3	4
	3	3	3	4	2	2	3	3	3	3	3	4	4	4	4	4	4	4	4	2	3	4
3	1	1	1	1	1	1	1	1	1	1	2	3	3	3	4	4	4	1	1	1	1	1
	2	2	2	3	1	1	1	1	1	2	4	4	4	4	4	4	3	3	3	1	1	1
	3	2	2	3	1	1	1	2	3	3	4	4	4	4	4	4	4	4	4	1	1	1
4	1	2	3	3	2	2	3	2	2	3	4	4	4	4	4	4	4	4	4	2	3	4
	2	3	3	4	2	3	4	3	3	4	4	4	4	4	4	4	4	4	4	2	3	4
	3	4	4	4	2	3	4	3	3	4	4	4	4	4	4	4	4	4	4	2	3	4

ACTION CATEGORIES

1 = no corrective measures
2 = corrective measures in the near future
3 = corrective measures as soon as possible
4 = corrective measures immediately

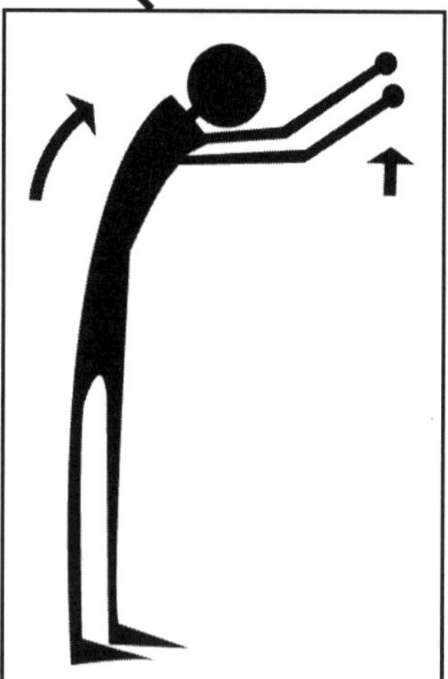

Source: see fig. 2.4.

Table 2.4   OWAS method: Action categories for work postures

BACK	1 straight	1	1	1	1	1	1	1	1	1	1
	2 bent forward	1	1	1	2	2	2	2	2	3	3
	3 twisted	1	1	2	2	2	3	3	3	3	3
	4 bent and twisted	1 2	2	2	3	3	3	3	4	4	4
ARMS	1 both arms below shoulder level	1	1	1	1	1	1	1	1	1	1
	2 one arm at or above shoulder level	1	1	1	2	2	2	2	2	3	3
	3 both arms at or above shoulder level	1	1	2	2	2	2	2	3	3	3
LEGS	1 sitting	1	1	1	1	1	1	1	1	1	2
	2 standing with both legs straight	1	1	1	1	1	1	1	1	2	2
	3 standing with one leg straight	1	1	1	2	2	2	2	2	3	3
	4 both knees bent	1 2	2	2	3	3	3	3	4	4	4
	5 one knee bent	1 2	2	2	3	3	3	3	4	4	4
	6 kneeling	1	1	2	2	2	3	3	3	3	3
	7 walking	1	1	1	1	1	1	1	1	2	2

% of working time   0   20   40   60   80   100

ACTION CATEGORIES

1 = no corrective measures
2 = corrective measures in the near future
3 = corrective measures as soon as possible
4 = corrective measures immediately

Source: see fig. 2.4.

Case-study 2A illustrates how the harmful action categories may be combined and presented to show the need for change. Clearly a follow-up study after change could be used to provide data to demonstrate any improvements. It also illustrates a participative approach to find the best solutions and their rapid acceptance and introduction.

CASE-STUDY 2A

Construction work

In construction work, the OWAS method has been used as a method to support and expand workplace surveys. 6,457 postures across 39 tasks and 12 jobs (including bricklayer, concrete layer, electrician, plumber, timberman, etc.) were studied. Inter-rate reliability coefficients were 85 per cent to 94 per cent. The proportions of harmful (*action category 3*) and very harmful (*action category 4*) work postures or work posture combinations in the tasks ranged from 2 per cent (electrician) to 28 per cent (cementer) of the working time. These are shown in the figure.

Figure: Percentage of harmful (action category 3) or very harmful (action category 4) work postures in construction work.

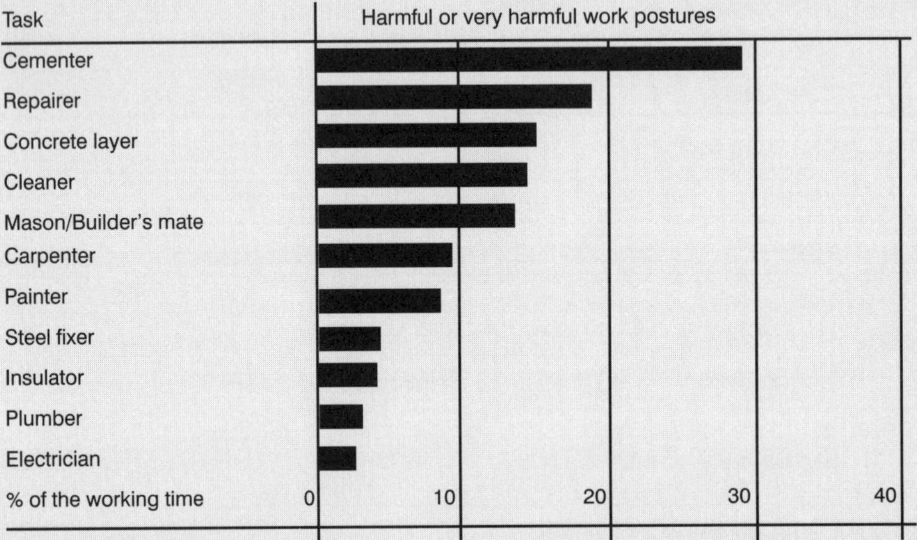

The team to improve the work comprised a safety manager, an occupational safety supervisor, an occupational health physician, an occupational health nurse, and a researcher. The team discussed measures to redesign the work so that the number of poor work postures could be reduced. Video recordings of the tasks were used as an auxiliary. All results obtained through computer analysis were discussed by health and safety personnel, ergonomists and workers' representatives. Working groups were then set up to alleviate any problems identified. The most important findings were:

- the practical, simple, clear procedures of OWAS were very appropriate to study in the construction industry;

- worker involvement is crucial in achieving acceptance and understanding of the programme;

- teamwork and group problem-solving are crucial to successful intervention; management, employees, safety experts and health professionals should be represented;

- feedback to employees at all stages is necessary to drive a continuous improvement process;

- implementation of change must be fast to sustain interest in the programme.

According to the researchers, the OWAS method produces information that motivates activity and cooperation at a workplace.

Source: Kivi and Mattila, 1991.

Mental work

Many types of work activity are indicated by the general term "mental work". No work, of course, is entirely physical or mental in nature; it all must involve at least some body movements and some interpretation of information and some thinking and decision-making. None the less, across much of the world there are changes under way, from factory to office work, from manufacturing to service industry, from active jobs with physical tools and machines to more sedentary jobs where it is information which is handled.

Just as no two jobs involving physical work are the same, so there is a vast range of jobs which are primarily mental in nature. These range from vigilance work as a monitor of occasional events (inspection tasks, or radar operations, or some continuous process control operations), to work in which novel problems must constantly be dealt with (some management work, professional work, many advanced technical jobs), to work primarily involving communication (training personnel, receptionist or telephonist). Keyboard-based work, although it appears to be an information handling job, is more akin to physical work, especially when it is data or text entry. Maintenance jobs can require a mix of diagnostic (mental) and rectification (physical) work.

As an equivalent of physical workload, jobs can entail a mental workload on people; performance will deteriorate with overload or underload. The operator who simultaneously has to attend to several displays of information, communicate with other operators or a supervisor, remember non-routine procedures and make decisions will make mistakes or miss or misinterpret important information. At the other end of the spectrum, monotonous and repetitive work, where the need for attention to what is done can be quite low, will also be prone to errors being made, due to operator lapses in concentration, to boredom or fatigue. Many inspection tasks, data-entry work, or any work which involves monitoring infrequent signals fall into this category.

Principles which can aid the design of work which is largely mental in nature, in particular for control of hazardous continuous processes, are shown in box 2.3.

Box 2.3   Checklist for design of control room operators' work in nuclear power plants

TASK DESIGN AND JOB ORGANIZATION

1. Operator performance

The operator's choice of behaviour should be clear and unambiguous.

- Are criteria for the choice of operator behaviour clear?
   The operator should be fully aware of targets, priorities and penalties for failure.

- Are there no conflicts and incompatibilities in these requirements?
   The operator should not be expected to resolve conflicts between productivity and safety, for example.

- Is there an attitude of non-penalization when an operator says there is a failure and there is not?
   This is especially important if the penalties for missing an equipment fault are high.

- Are the criteria clear and unambiguous for taking over manual control from automatics?

2. Operator assistance

The operator should be given help in coping with high or multiple workloads.

- Can several different tasks, all the responsibility of one operator, be left to run unattended for a while, so that attention can be given to the task which needs attention?

- Are visual or audible alarms given for those situations requiring prompt action?

- Are there aids to help the operators to find their place in an operating sequence if they have been doing another task or are otherwise interrupted?

- Does the operator have time to find out what is happening, and think, before being expected to take over manual control from automatics?

- Are high speed, high accuracy or highly repetitive tasks done by automatic control?

3. Optimum operator activity

The work done by the operators should give each of them a feeling of personal responsibility.

Box 2.3   (cont.)

35

4. Team responsibility

The allocation of information and responsibility within teams of workers should be clear.

- **In teams, is the allocation of responsibility and authority clear, complete, non-overlapping, known to and accepted by the operators (for example, when several people work on the same control panel)?**

- **Are the changes in these responsibilities during an emergency clear and practised?**

- **Are there clear procedures for the handover of information and responsibility between:**

  (a) different shifts?
  (b) people with different responsibilities (for example, operators and maintenance)?

- **Is the social situation such that an operator can indicate the errors of others, including superiors, without embarrassment?**

5. Operators' individual responsibilities

The operator's level of activity should be kept at an optimum.

- **Are mental and physical workloads at a level which can be maintained without strain for several hours, or only up to high levels for short periods with recuperation time?**

- **Are periods of sustained concentration shorter than one hour? This also applies during emergencies.**

Tasks should be allocated to the operators to maintain their alertness, skill, interest and self-respect, even where this is not technically necessary.

- **Do the tasks given to one worker make a meaningful combination? If not, a good explanation should be given of why tasks are grouped together.**

- **Do the tasks give the operator a feeling of personal value, and of responsibility for plant safety and for the quality of the product?**

6. Incident follow-up

There should be clear mechanisms for following up incidents.

- **Is there a confidential near-incident or incident reporting system?**

- **Is the information obtained from incidents on this and other similar plants used to modify the plant and the designs of such future plants?**

- **Is the information obtained from incidents used in retraining operators?**

- **Is the information fed back to the original design engineers?**

Source: United Kingdom, Atomic Energy Authority, 1985.

Stress

Many of the effects of badly designed work, whether physical or mental, will be felt by individuals in terms of occupational stress caused by not coping with the workload. This may be so particularly for mental work. Many different approaches have been taken to the understanding, measurement and alleviation of stress or stress effects.

How does stress arise?

When people are faced with demands from others or from the physical or psychosocial environment to which they feel unable to respond adequately, a response of the organism is activated to cope with the situation. The nature of this response will depend upon a combination of different elements, including the extent of the demand, the personal characteristics and coping resources of the person, the constraints on the person in trying to cope and the support received from others.

Under normal circumstances people should be able, by activating their reaction mechanisms, to find new balances and responses to new situations. Stress is therefore not necessarily a negative phenomenon. It would be a mistake to concentrate only on the pathological aspect of stress without emphasizing its importance in the search for dynamic adaptation to a given situation. If health is considered as a dynamic equilibrium, stress is part of it, for there is no health without interaction with other people and with the environment. Only excesses are pathological.

Some stress, then, is normal and necessary. But if stress is intense, continuous or repeated, if the person is unable to cope or if support is lacking, then stress becomes a negative phenomenon leading to physical illness and psychological disorders. From early disorders to real illness, the harmful consequences of stress cover a broad range from chronic fatigue to depression, by way of insomnia, anxiety, migraines, emotional upsets, stomach ulcers, allergies, skin disorders, lumbago and rheumatic attacks, and tobacco and alcohol abuse, and can culminate in the most serious consequences of all: heart attacks, accidents and even suicides.

Negative stress has many causes. Some of these are to be found in an unsatisfactory fit between the individual and the physical environment. Stressors of this type relate to noise, odours, illumination, temperature, humidity, vibrations, crowding, dangerous substances, machines and tools. Other stresses are generated primarily by the relation between individuals and their psychosocial environment. These can depend on the level of autonomy and responsibility, the load of activities, the organization of different activities, the arrangement of working time, the relationship with other individuals and communities, and so on. Reference is often made to physical stress, on the one hand, and psychosocial stress, on the other, although they are so interlinked that a real separation is almost impossible.

The notion of stress thus challenges traditional categories because it bridges physical, mental and social well-being.

Altogether stress is generally experienced as a result of a discrepancy between people's perception of the demands on them and their perception of their ability to cope with the demands. Stress felt will be modified by constraints, support and individual characteristics (figure 2.6).

Figure 2.6   Factors involved in the perception and experience of stress

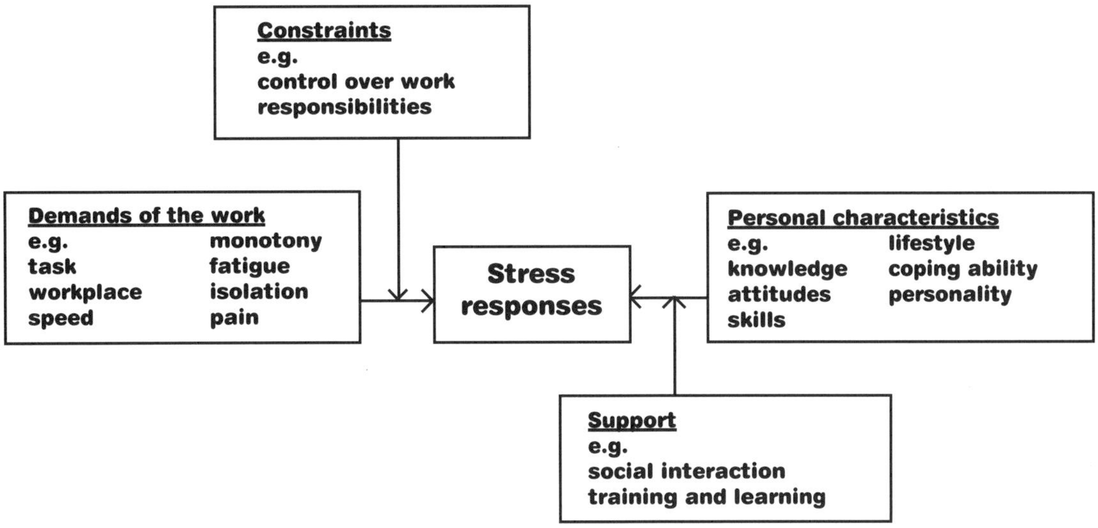

Stress recognition and assessment

Preliminary to any anti-stress intervention is the recognition that stress "exists" in the specific workplace and the identification of the combination of factors that generate it. For effective action to take place, it is essential that the recognition of the signs and symptoms of stress take place as early as possible. Even though each of the signs and symptoms of stress may be due to other factors, the occurrence of several of these signs and symptoms at once may require that anti-stress action be taken.

At the individual level, the following physical, behavioural, mental and emotional signs may be apparent:

- dry throat, muscle tension, headaches, indigestion, tics, insomnia, high blood pressure, etc.;

- irritability, impulsive behaviour, difficulty in making decisions, sudden increase in smoking or alcohol use, etc.;

- excessive worrying, feeling of worthlessness, brooding, forgetfulness, being easily startled, day-dreaming, etc.

At the workplace level, high levels of absenteeism, turnover, accidents at work (including minor accidents) and disabilities are often linked with stressful situations. Identification of the stressors causing high-stress situations is then required.

Involving workers in identifying those stressors which they feel cause unnecessary stress in their jobs, and in rating them to establish priorities for intervention, is highly recommended. The assessment should be done in a systematic way, and employees should be asked to express their concern about any situation that may be causing stress at work.

In this respect, recourse to an "audit" of the relevant hazards can prove particularly helpful, as exemplified in the following checklist (table 2.5).

Table 2.5 Checklist of stress hazards

Work characteristic	Hazardous conditions (High likelihood conditions)	Absent/Low <u>or</u> Present/Medium <u>or</u> Obvious/Severe (please specify)
Organizational function and culture	Poor communications Organization as poor task environment Poor problem-solving environment Poor development environment	
Participation	Low participation in decision-making	
Career development and job status	Career uncertainty Career stagnation Poor status work Work of low social value Poor pay Job insecurity or redundancy	
Role in organization	Role ambiguity: not clear on role Role conflict Responsibility for others or continual contact with other people	
Job content	Ill-defined work High uncertainty Lack of variety Fragmented work Meaningless work Underutilization of skill Physical constraint	
Workload and work pace	Work overload Work underload High levels of pacing Lack of control over pacing Time pressure and deadlines	
Working time	Inflexible work schedule Unpredictable hours Long hours or unsocial hours Shiftworking	
Interpersonal relationships at work	Social or physical isolation Lack of social support from other staff Conflict with other staff Violence Poor relationships with supervisors and managers	
Home/work interface	Conflicting demands of work and home Low social or practical support from home Dual career problems	
Preparation and training	Inadequate preparation for dealing with more difficult aspects of job Concern about technical knowledge and skill	
Other problems	Lack of resources and staff shortages Poor work environment (lighting, noise, bad postures)	
Source: Cox, Griffiths and Cox, 1996.		

Anti-stress intervention

Once the existence of stress has been recognized and the stressors identified, action to deal with stress should be taken. Assuming that stress is a misfit between the demands of the environment and the individual's abilities, the imbalance may be corrected according to the situation, either by adjusting external demands to fit the individual or by strengthening the individual's ability to cope, or both. At this point, it should be kept in mind that since stress is a multifaceted phenomenon, no simple solution is available. Furthermore, it is impossible to provide a unique solution to manage stress because of the particular circumstances of each case.

It is generally recognized that the ideal way to combat stress would be to prevent its occurrence. This might be achieved by tackling the core of the problem – the cause. However, there is no single cause of stress and the elimination of all stressors is a utopian task. Therefore, action should be aimed at eliminating as many causes as possible, so that the action taken reduces stress and prevents future stress. As this cannot always be achieved in the short term, it is generally agreed that improving the ability to cope with stress is a valuable strategy in the process of combating stress.

The following is a possible list of types of intervention, ranging from those on the environment to those targeted to the individuals.

Intervention of the external socio-economic environment
> Legislation, international and national directives
> Social support

Intervention on technology and work organization
> Improving job planning and reliability of the work systems
> Reduction of working times and arrangement of working teams and rest pauses in relation to the workload
> Arrangement of shift schedules according to psycho-physiological and social criteria
> Participation in decision-making

Intervention in the workplace and task structure
> Improving the work environment
>> Lighting
>> Noise
>> Microclimatic conditions and indoor air quality
>
> Arranging workplaces according to ergonomic criteria
>> Workstation design
>> Working with visual displays units
>> Sitting postures

Intervention to improve individual responses and behaviour
> Individual ways of coping with stress
> Selection and training
> Counselling and other supporting measures at company level

Specific intervention for health protection and promotion
> Appropriate medical surveillance

The effectiveness of a preventive approach in combating stress while improving enterprise efficiency is shown in the following case-study.

CASE-STUDY 2B

Anti-stress programmes
For managers, supervisors, operators, loaders, foremen and maintenance staff at Western Coalfield Limited in India

As a result of this training programme, based on the reclassification of tasks and group dynamics, more than 50 per cent of operators and loaders indicated a reduction of monotony and boredom and improvements in their working schedules. Between 25 and 50 per cent indicated improved responses to stresses in the physical environment, significant relief from physical ailments and a reduction of smoking.

For postal workers in the United Kingdom

Among the workers who benefited from stress counselling, anxiety was reduced by 29 per cent, psychosomatic disturbances by 40 per cent, and absence for sickness by 46 per cent in terms of the number of absences and 60 per cent in terms of the number of days.

For blue-collar workers at ABB in Sweden

By introducing an anti-stress programme based on job enrichment and job rotation, turnover was reduced in the area of intervention from 39 per cent to 0 per cent. Absences for sickness, which had totalled as much as 35 per cent in 1989 (one day lost for every three working days) were considerably reduced. This reduction affected primarily workers without chronic health problems, bringing their rate of absenteeism down from 14 to 2 per cent. Productivity increased. "On-time" deliveries, for instance, increased by 10 per cent to 98 per cent.

For managers and assembly-line workers in a multinational manufacturing company in Mexico

Based on an extensive involvement of all parties concerned in analysing and shaping work organization and individual responsibilities, this programme resulted, after 12 months, in a decrease in psychosomatic illness of 15 per cent, in a decrease in absenteeism and an increase in productivity. The general human and work environment in the company improved substantially.

Source: ILO, 1992.

Job design

It is with people's roles at work, the ways in which tasks and other responsibilities are combined, that job design is concerned. The need to find a better way to motivate people to perform well and to give them job satisfaction has long been felt and discussed. Even if they ever were adequate, the so-called Taylorist traditional approaches of work simplification, measurement and control, dating from the early years of the century, are certainly not appropriate for the needs of industry today.

Later, in Chapter 5, we will talk about the various ways of arranging and organizing work to make it more of a worthwhile activity for the people doing it. Here, some of the ideas which have been developed over the last three or four decades concerning the needs which people today expect to satisfy in their jobs will be presented. Matching these needs is an important aspect of job design.

Box 2.4 indicates some basic work-related needs which, if not satisfied in some measure, will limit people's attachment and motivation to their jobs.

Box 2.4   Basic work-related needs relevant to job design

1. **Job content that is reasonably demanding (other than sheer endurance) and also provides variety.**

2. **Long-cycle self-paced work, with the opportunity to move around physically.**

3. **Ability to learn new skills, have some freedom in method of working and few time pressures (e.g. tight deadlines).**

4. **Decision-making, including planning and control of work and opportunities to exercise discretion.**

5. **Social support, interaction with colleagues and recognition of contributions in the workplace.**

6. **Feelings that the job leads to some sort of desirable future; a positive work management climate; participation in decisions which affect their work life.**

7. **Ability to relate what they do and what they produce to the life of the wider community, and to feel that they make a worthwhile contribution to it.**

Good job design depends upon finding out what **job characteristics** are important to the relevant group and then designing work to improve or increase them. Examples of job characteristics are autonomy, feedback on performance, clarity of role, variety of tasks and utilization of skills. Figure 2.7 summarizes one model of doing this, the Job Characteristics Model.

Another overall framework of what should be aimed for in job redesign is shown in table 2.6, based on understanding and implementing the optimum combinations of control, demands, responsibilities, support and interaction for any set of work tasks and group of workers. It is by using the knowledge and ideas of those workers for guidance, that most appropriate job redesigns will be achieved.

Figure 2.7   The job characteristics model

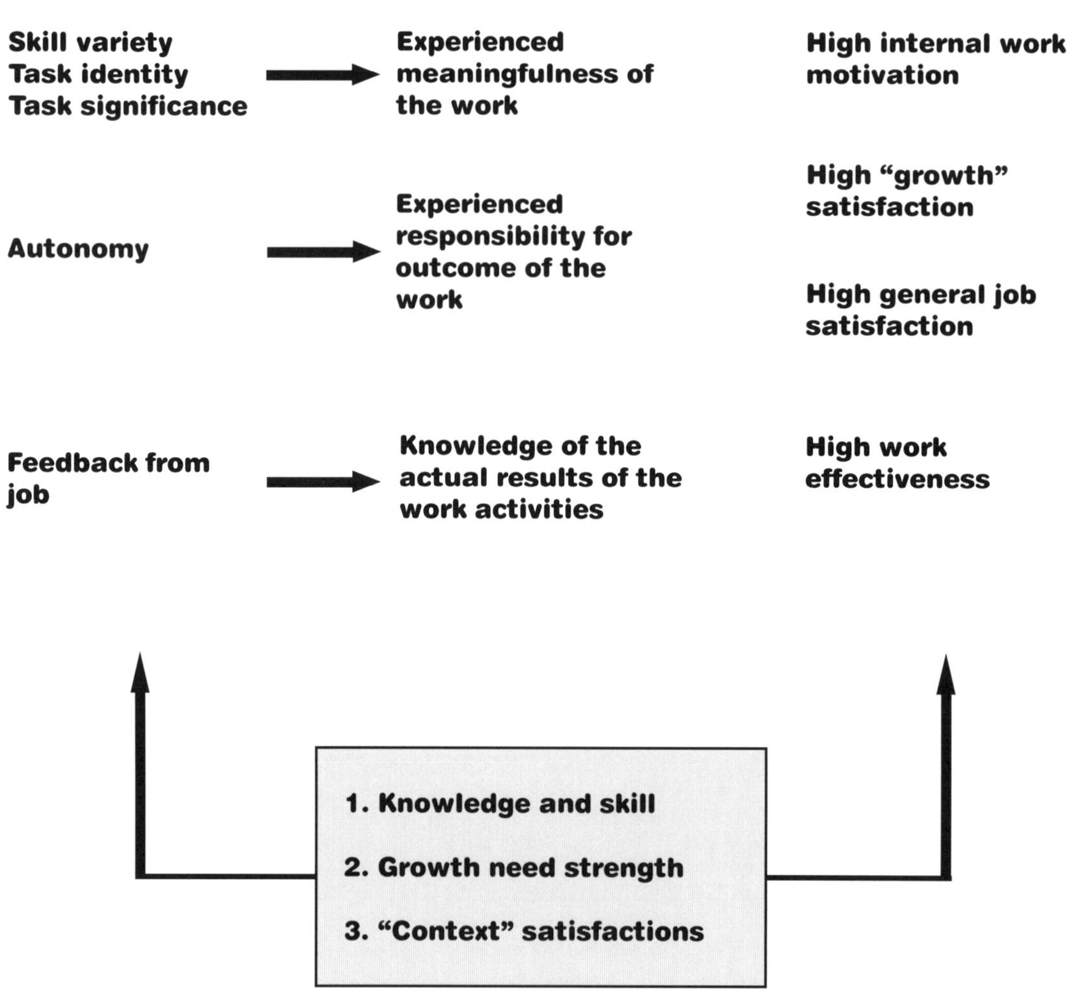

Source: Hackman and Oldham, 1980.

Table 2.6 A framework for job redesign: Variables and propositions

Part 1. Categories of independent variables with outline definitions

Category	Variable	Outline definition
Control	Timing control	Individual control over when to complete tasks: degree to which operator must respond as and when technology requires.
	Method control	Individual control over how to complete tasks: degree to which technology (or other factors) confines choice.
	Boundary control	Range and scope of individual to pay close and constant attention.
Cognitive demand	Monitoring demand	Requirement for the individual to pay close and constant attention.
	Problem-solving demand	Requirement for the individual to diagnose and solve problems.
Responsibility	Production responsibility	Responsibility for technology and output (e.g. likelihood of the individual making expensive errors).
Social interaction	Social support	Quality of relationship with others.

Part 2. Summary of the propositions

1. The greater the variance to be controlled at source, the greater will be the performance benefits arising from high-boundary control by operators.

2. Job-related strain results from the co-occurrence of high monitoring demand and high production responsibility.

3. Job-related strain results from the co-occurrence of high demands with low boundary control.

4. The job-related strain effects of monitoring demand, production responsibility and control are accentuated by low social support, and reduced by high social support.

5. Job satisfaction increases as an additive function of greater timing, method and boundary control, lower monitoring demand, and higher problem-solving demand, production responsibility, social contact and social support.

Source: Wall et al., 1990.

Management programmes for job redesign

Since prevention is always the best management strategy, it is best to pick up the earliest signs of problems arising and institute a programme of change. The major points in a monitoring and intervention programme are given in table 2.7.

Table 2.7   Monitoring and intervention programme

General requirements	Specific intervention sequence
Encourage management-employee cooperation in good work conditions by information and education programmes	(1) Data collection (medical and personnel records; operator surveys and interviews; workplace observations; video/photos for records and analysis)
Early reporting and intervention programmes continuing during life of the company	(2) Analysis of data
Information from medical, personnel, safety, training, shop management and workforce	(3) Discussions with relevant people, including operators
Safety and industrial engineering functions monitor for relevant problems	(4) Plan of action, with priorities and completion dates
Intervention sequence as in right-hand column	(5) Implementation, using mock-ups to test, if needed
Regular report and review by senior management	(6) Monitoring and modification
	(7) Assessment of effectiveness and publication of results

The approaches in this programme are threefold:

1.   *Education and awareness:*

Management and employees need reliable information about likely problems, and what they must do if they suspect that they exist. Information about any particular monitoring or intervention programme should be widely known to encourage cooperation.

2.   *Early reporting and intervention:*

Mechanisms to facilitate fast and appropriate action if a problem is revealed. The programme seeks to achieve effective reporting mechanisms and interdepartmental communication for pooling of skills and knowledge.

3.   *Assessment and specific intervention:*

This part has four main aims:

(i) to discover the extent of the revealed problem;

Medical and personnel records are surveyed to assess the extent of any personal injuries, (e.g. work-related upper limb disorders) arising from the problem, to isolate which jobs are showing problems, how long those who have been sick or absent have worked in the problem area, how much absence is involved, who are the people involved and where any symptoms are being revealed.

These data are then used to prioritize problem areas for specific intervention; the most serious ones being those where injury occurs quickly and in the largest numbers.

(ii) to investigate sites where the problem is occurring;

At a site chosen for intervention, information is collected and is assessed using a range of methods:

- workplace measurements (including tools and equipment design)

- posture recording (e.g. OWAS, rapid upper limb assessment procedure, photographs, videos);

- body part discomfort rating;

- work information (forces, repetition, etc.);

(iii) to identify probable causes;

Data collected are analysed through two broad approaches:

- measurements and other information collected are compared to published data or recommendations;

- responses, e.g. discomforts and observations, concerning loads imposed by the work are interpreted in terms of mismatches between the person and the work.

A list of all the possible causes of problems is made; for example, how do the workplace and operator dimensions compare, are the body, arm, shoulder and/or neck positions explained by incorrect bench or seat height, or machine obstruction, poor positioning, etc.?

(iv) to produce feasible solutions and instigate and monitor effective action;

Possible improvements are listed in priority order which gives highest places to frequent repetitions, forces, high body part discomfort values or other serious issues. These form the basis for discussion with all those involved and an action list is produced.

Changes may then be introduced; when redesigning workplaces, advice is given in Chapter 3 on workstations design in this manual. One point is of note: it is better to test a design via a mock-up, letting people of different sizes run through the work sequence and find out any problems, rather than build a new design straight from the drawing-board and find that it does not suit some employees. Changes are monitored and their effectiveness assessed using the same methods as previously to allow direct comparisons to be made.

It is essential that people are involved in both the investigation and in the introduction of change. Then not only are the data relating to the problem more accurate and reliable, but the change will be easier and quicker to introduce (see case-study 2C).

CASE-STUDY 2C

Electronics assembly

A Swiss electronics company, most of whose assembly workers are "unskilled" immigrant women, wanted to redesign jobs in printed circuit board (PCB) assembly to provide attractive jobs, increase productivity and improve flexibility. The 100 workers and supervisors were treated by the external consultants as "experts".

Employees were first introduced to the idea of participation and the project goals. Two "project groups" were formed from each of the 15-20 most interested employees, and were given the technical goals for the system. During a long process of planning and preparation, the groups met often for two to four hours during a working day or sometimes in restaurants for a full day off-site. A "metaplan" technique was used: all ideas and problems were recorded on coloured cards and grouped on large pin boards around the room; open discussion, interpretation and regrouping of all ideas could then take place. At the end of each meeting, all card groupings were photographed and photocopies distributed as a protocol record to group members or to others.

The principal change implemented as a result of the process was of self-managed teams with responsibilities for complete assembly of a product, non-electronic testing, internal task distribution and delivery of products. Former supervisors became "advisors" for trouble-shooting and instruction. Continual logging and discussion of problems took place after implementation.

Key issues identified by the consultants were that the process had to deal with under-educated workers often with language barriers (group discussions were held in various languages!); the women seemed to increase in self-confidence (possibly causing them domestic difficulties); the advisors' roles were crucial and difficult; mechanisms had to be set up to deal with envy and inter- and intra-group competition, including peer pressure on group members to speed up production; and every case will be a one-off, requiring self-design by the "local experts".

Source: Frei et al., 1993.

References

D.C. Berliner, D. Angell and J.W. Shearer: *Behaviours, measures and instruments for performance evaluation in simulated environments*, paper presented to the Symposium on Quantification of Human Performance, Albuquerque, New Mexico, 1965.

R. Birnbaum, A. Cockroft, B. Richardson and E.N. Corlett: *Safer handling of loads at work: A practical ergonomics guide* (Nottingham, Institute for Occupational Ergonomics, 1992).

T.S. Clark and E.N. Corlett: *The ergonomics of workspaces and machines* (London, Taylor & Francis, 1984).

C.W. Clegg: "The derivation of job designs", in *Journal of Occupational Behaviour*, Vol. 5, 1984, pp. 131-146.

E.N. Corlett: "Static muscle loading and the evaluation of posture", in Wilson and Corlett (eds.), 1989.

—: "Ergonomics fieldwork: An action programme and some methods", in M. Kumachiro and E.D. Megaw (eds.): *Towards human work* (London, Taylor & Francis, 1991).

T. Cox: "The recognition and measurement of stress: Conceptual and methodological issues", in Wilson and Corlett (eds.), 1989.

—, A. Griffiths and S. Cox: *Work-related stress in nursing: Controlling the risk to health*, CONDI/T/WP.4/1996 (Geneva, ILO, 1996).

F. Daniellou, A. Garrigou, A. Kerguelen and A. Laville: "Taking future activity into account at the design stage", in C. Haslegrave et al. (eds.): *Work design in practice* (London, Taylor & Francis, 1990).

G.C. Drury: "Task analysis methods in industry", in *Applied Ergonomics*, Vol. 14, 1983, pp. 19-28.

J.A.E. Eklund: "Organization of assembly work: Recent Swedish examples", in E.D. Megaw (ed.): *Contemporary ergonomics 1988* (London, Taylor & Francis, 1988).

D.P. Ferreira and M.F. Tracy: "Musculo-skeletal disorders in a brick company", in E.J. Lovesey (ed.): *Contemporary ergonomics 1991* (London, Taylor & Francis, 1991).

F. Frei, M. Hugentobler, W. Duell, A. Alioth, and S.J. Schurman: *Work design for the competent organization.* (Westport, Connecticut, Quorum Books, 1993).

E. Grandjean: *Fitting the task to the man* (London, Taylor & Francis, 1982).

S.M. Grey, B.J. Norris, and J.R. Wilson: *Ergonomics in the electronic retail environment* (Slough, United Kingdom, ICL, 1987).

J.R. Hackman and G.R. Oldham: *Work redesign* (Reading, Massachusetts, Addison Wesley, 1980).

P. Hellstrand: *New job content produces good results* (Stockholm, Swedish Work Environment Fund, 1989).

ILO: *Conditions of Work Digest* on *Preventing stress at work*, Vol. 11, No. 2, 1992.

Institution of Production Engineers: *Design for better work* (London, 1986).

G. Kanawaty: *Introduction to work study* (Geneva, ILO, fourth revised edition, 1992).

P. Kivi and M. Mattila: "Analysis and improvement of work postures in the building industry: Application of the computerized OWAS method", in *Applied Ergonomics*, Vol. 22, 1991, pp. 43-48.

V. Louhevaara and T. Suurnäkki: *OWAS: A method for the evaluation of postural load during work*, Training Publication No. 11 (Helsinki, Institute of Occupational Health, 1992), pp. 4, 9 and 10.

L. McAtamney and N. Corlett: "Work-related upper limb disorders", in *Applied Ergonomics*, Vol. 24, No. 2, 1993, pp. 91-99.

D. Meister: *Behavioural analysis and measurement methods* (Chichester, John Wiley, 1985).

R.B. Stammers and M.S. Carey: "Task analysis", in J.R. Wilson and E.N. Corlett (eds.), 1989.

M.G. Stevenson and K.N. Baidya: "Some guidelines on repetitive work design to reduce the dangers of tenosynovitis", in M.G. Stevenson (ed.): *Readings in RSI. The ergonomics approach to repetition strain injuries* (Sydney, NSW University Press, 1987).

Swedish Work Environment Fund: *Stress, health and job satisfaction* (Stockholm, 1989).

United Kingdom Atomic Energy Authority. Safety and Reliability Directorate: *A guide to reducing human error in process operations: Short version*, document SRD-R347 (London, 1985).

United Kingdom. Health and Safety Commission: "Manual handling operations regulations" in *Guidance on regulations*, Legal No. 23 (London, HMSO, 1992).

United Kingdom. Ministry of Defence: *Naval engineering standards: Human factors in design of military-computer systems* (London, HMSO, 1986).

P. Van Wely: "Design and disease", in *Applied Ergonomics*, Vol. 1, 1970, pp. 262-269.

T.D. Wall, J.M. Corbett, C.W. Clegg, P.R. Jackson and R. Martin: "Advanced manufacturing technology and work design: Towards a theoretical framework", in *Journal of Organizational Behaviour*, Vol. 11, 1990, pp. 201-219.

—, K. Davids and P.R. Jackson: *Job attitude questionnaire*, unpublished (Sheffield, University of Sheffield, 1991).

J.R. Wilson: "Framework for ergonomics methodology", in J.R. Wilson and E.N. Corlett (eds.), 1989.

— and E.N. Corlett (eds.): *Evaluation of human work*, 2nd edition (London, Taylor & Francis, 1995).

WORKPLACE DESIGN

3

John Wilson

Designing workplaces

The tasks that people perform, the jobs and roles they hold, and the machines and interfaces they use do not exist in a vacuum. How well – how effectively, safely, healthily, satisfyingly – these things are done will be affected by the next level of interaction in any human-machine system; how well people *fit* with their physical workspace and physical work environment.

Major issues for provision of healthy and efficient workplaces and environments are:

- **worker task position:**   reach and grasp distances and orientation
  working zones
  lines of sight
  work heights

- **posture:**   all facets of working position
  seated and standing work
  furniture and equipment design

- **clearances:**   access and fit
  movement space
  activity space

- **machine control:**   control and handle dimensions, clearances, visibility

- **force application:**   allowable forces in certain postures
  material handling

- **workstation layout:**   display and control positions
  display-control relationships

- **physical environment:**   lighting, noise, climate, vibration, radiation,
  chemical, psycho-social, spatial, etc.

Workplaces and the physical environment comprise what is probably the best opportunity to apply a participative methodology in the assessment and redesign of work. Experience shows that people are most willing and able to contribute to workplace design; issues usually become clearer and the chances of producing successful solutions greater. In general, the most successful workplace changes will be achieved through application of participative practices.

Importance of appropriate work environments

Different aspects of the environment have positive and negative effects on all areas of individual and company well-being (figure 3.1). All environments at the extreme will affect people's health; what we need are data to tell us at what levels such effects as increased errors or accidents, noise-induced hearing loss, heat stress, or musculo-skeletal diseases might be found. We need to know not just the physical stimulus measurements, but also exposure times, and also what alleviating effect different job, task or equipment designs, or personnel selection, might have.

Ideally we are concerned with keeping harmful environmental variables at well below the levels where health or poor performance effects are found.

Therefore we wish to eliminate causes of annoyance, discomfort or dissatisfaction, such as perceived flicker from fluorescent lights, discomfort glare on a visual display unit, irritating air draughts, noise annoyance, cramped workspaces, etc.

Figure 3.1 Effects of the working environment

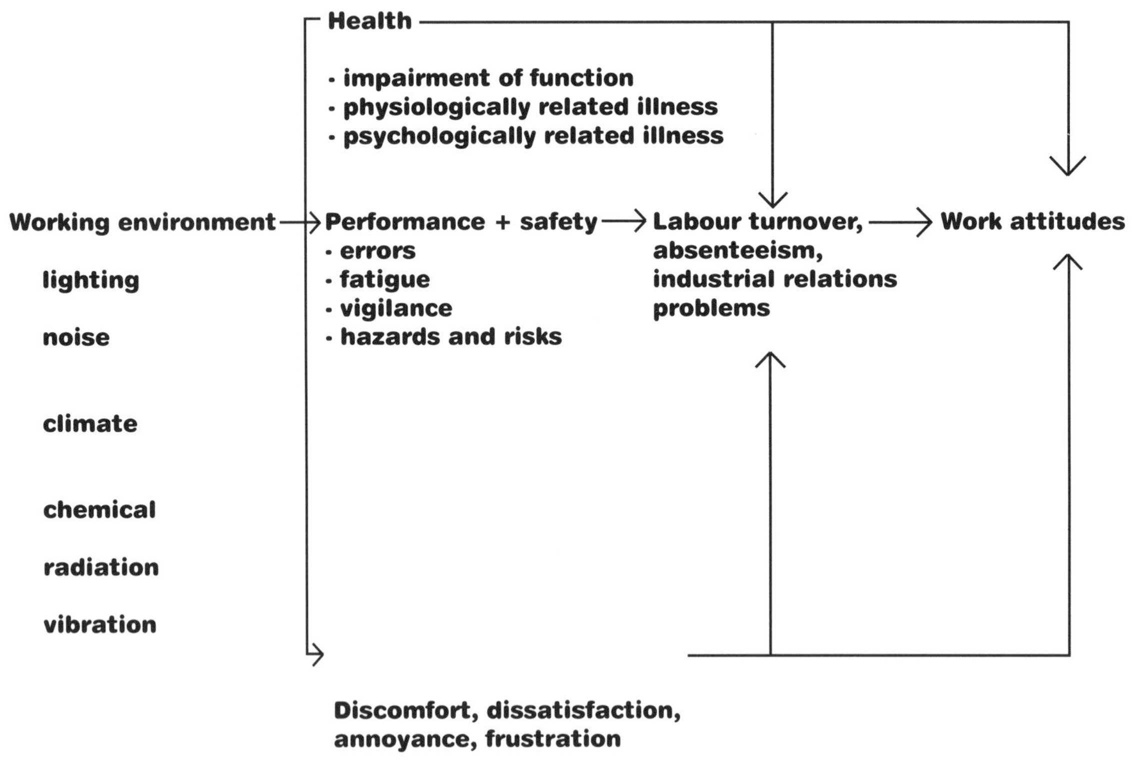

Health
- impairment of function
- physiologically related illness
- psychologically related illness

Working environment
lighting
noise
climate
chemical
radiation
vibration

Performance + safety
- errors
- fatigue
- vigilance
- hazards and risks

Labour turnover, absenteeism, industrial relations problems

Work attitudes

Discomfort, dissatisfaction, annoyance, frustration

Source: Adapted from Wilson and Grey, 1986.

Another major effect of the work environment is upon performance. This can occur indirectly through health problems or discomfort and dissatisfaction, but may also be produced more directly. Examples are glare, making it difficult to see the screen on a VDU, noise levels which interfere with hearing speech, cold conditions leading to loss of dexterity, or reduced vigilance in inspection through distraction from poor seating. Performance reductions may be found in terms of output and errors for both mental and physical tasks, and with respect to both direct task-related measures and systemic ones, such as absenteeism or labour turnover.

A further result of working environments, good or poor, is the attitudes that such conditions might engender in the workforce. This might be seen in, for instance, such opinions (spoken or unspoken) as "If management thinks so little of us to give us such conditions to work in, then why should we cooperate with what they want?". Consequences will be resistance to change, lack of innovation or dynamism, a tendency to cure rather than prevention, and generally an unwillingness to give of skills and time. Working environments perceived as good, on the other hand, may evoke opposite reactions.

Process for investigating and designing workplaces

Summarized below are the main stages of any workplace redesign process; in practice some of these stages may be unnecessary or impractical and the process will certainly not run in a smooth sequential fashion. At the least, there will be iterations back to earlier stages if problems crop up later.

Analysis phase

It is important at the outset to clearly define objectives – both for the redesigned or new workplace and for the ergonomics investigation and specification. Consultation with all the workforce affected – management, service departments and operators – should take place very early on and participative procedures setup (see Chapter 5). Existing records – for instance, of accidents, absenteeism, machine down-time, health complaints, etc. – will yield valuable preliminary analysis data; new data may also be collected. If we are redesigning an existing workplace, this is the time to observe what is happening now – what tasks are being performed, how, by whom, where, when and why. Direct observations may be employed – by taking videos or photographs or by the observer writing and drawing descriptions, or by use of an observation checklist. Indirect observations are made by conducting interviews with, or giving questionnaires and rating forms to, relevant staff. Examples of techniques for observation and recording are also given in Chapter 2.

The purposes of the observations are to identify key ergonomics problems in the work and workplace, to feed into a task analysis, and to provide some benchmark data against which to compare the new design.

Development phase

In development of the improved work and workplace, the task and what the operator has to do must be central. First, decisions must be made about what functions will be carried out by people (and by which people – *division of function*) and which will be machine or computer tasks or automated tasks (*allocation of function*). Defining these tasks, and for workplace layout particularly the physical and monitoring ones, allows us to position them around the operator so as to minimize postural problems (task location); this is irrespective (for the moment) of what work furniture is required. In turn then, the workspace is laid out to make appropriate allowance for operators' physical fit and clearance between them and equipment or furniture; to define the workspace in the light of operators' needs of reach, movements and visibility; to arrange equipment position (displays and controls) according to criteria of priority, sequence of operation, function, safety or convenience; to plan the wider workspace to ensure that there is enough room for normal access and also to allow space required for maintenance activities; and, finally, to specify lighting, noise and climatic conditions. The design of the plant and facilities required to do this is described in the following chapters.

Evaluation phase

What appear to be ideal solutions at first may well not be feasible for technical or economic reasons, and compromises may have to be made. Checks also should be made with health and safety legislation, etc. Evaluations of the proposed workstation can take place in controlled conditions isolated from the work site (laboratory trials) or actually installed at the work site (field trials). Often a mixture of testing will be best, using laboratory trials to test out many small adaptations of a basic idea and the field trial to validate the solution under realistic conditions. The theme of evaluation should continue after system implementation, collecting observation, performance and user data periodically.

Sources of information

Information will be needed to assess existing workplaces and to redesign and evaluate new ones. Data requirements are summarized in table 3.1.

Table 3.1   Ergonomics design data and examples of their use

Data requirements	Examples of areas of application
I. Body dimensions (a) Body weight, height, length and circumference of body members. (b) Bending and crouching capabilities, reach and clearance dimensions. (c) Strength and static and dynamic force measurements; grip, lifting, pushing and pulling forces; body impact forces. (d) Balance and centres of mass; stability, loss of control, vestibular factors. (e) Changes in these dimensions as function of sex, age, etc.	Design of tools and other hard goods; development of clothing sizing. Clearances. Control location and operation. Workspace design. Equipment and job design for industrial workers. Product compatibility design.
II. Physiological and environmental (a) Physiological activities, such as muscle behaviour, blood flow, heart rate, fatigue, endurance, energy expenditure, peak load capacities. (b) Limits for, and tolerance to, thermal, visual, radiative, auditory, vibratory or chemical environments.	Work capacities and work activities, manning levels and schedules. Postural loadings. Environmental design, toxicity levels, evaluation of tasks in e.g. agriculture, logging, steel-making.
III. Psychological (a) Sensory processes: visual, auditory, tactile, proprioceptive. (b) Perceptual motor skills: learning, manipulation and control, speed and accuracy. (c) Information processing: reaction time, simultaneous inputs, overload, feedback, attention, uncertainty, interpretation of symbols, memory. (d) Attitudes and behaviour: motivation, risk-taking, activity patterns and stereotypes, decision-making consumer behaviour.	Design of controls and displays, digital and analogue; visual and auditory information and warning systems. Display-control relationships, machinery stopping time. Design of instructional and training materials.
IV. Special population capabilities Anthropometric, physiological and psychological measures of children, older persons, the disabled, etc.	Special equipment, building, product and environmental design.

Source: With acknowledgements to Ramsey, 1978, and Van Cott et al., 1978.

Recommendations on dimensions for population groups

It is impossible to summarize recommendations for physical workplace dimensions and values; what is appropriate for a bottle inspection plant will not be so for a train cab! What is more, the body dimensions of American people are quite different from the Chinese population, and even the proportions differ; for example, the trunk to leg lengths differ for different groups around the world, which make important distinctions if designing a seated workplace.

Many countries have now published anthropometric measurements for their populations and, where a population is a mixture of different populations, different sets of values exist. What is a common observation is that knowledge of one dimension does not allow us to predict the others, e.g. short people do not always have short arms, or tall people may have small hands. Hence, any decisions on workspaces should be based on real data, not extrapolation from a textbook table and measures only of the stature of local people. No general rule should be applied without thought, but as a guide, use people's maximum sizes for dimensions relevant to clearance (that is, design to fit the largest), and dimensions from the smallest sizes where reach distances are needed (i.e. design to fit the smallest).

Where information is not available, and even anthropometric tables do not always give the necessary figures, the information of figures 3.2(a) and (b) can be used. These are based on British population measures so will be acceptable for most populations of European origin. For other populations, as an approximate guide, the column with a standing height closest to that of the population under consideration can be used. This information can be augmented by measures on a sample of the shortest and tallest among this population so that clearance, reach, sitting and working height dimensions can be modified to suit local circumstances.

Sample user data

Instead of going to published sources of human dimensions, it is possible to collect relevant measures for the actual users from workers in the workplace. This is generally not preferred because of the time, effort and specialized expertise required to perform anthropometric or biomechanical survey work. Also, what is appropriate for today's work group may not be so for workers who join the company next year.

Collecting original physical data on a particular user or worker group is only justifiable in certain circumstances: when those data do not exist in the literature; when they exist for an irrelevant population group; when the workplace is to be tailored only to the requirements of that work group; or when the collection of such data is part of the motivational and learning process within a participative structured approach.

Figure 3.2　Dimensions for the adult population in the United Kingdom:
(a) Clearance and access for a standing worker

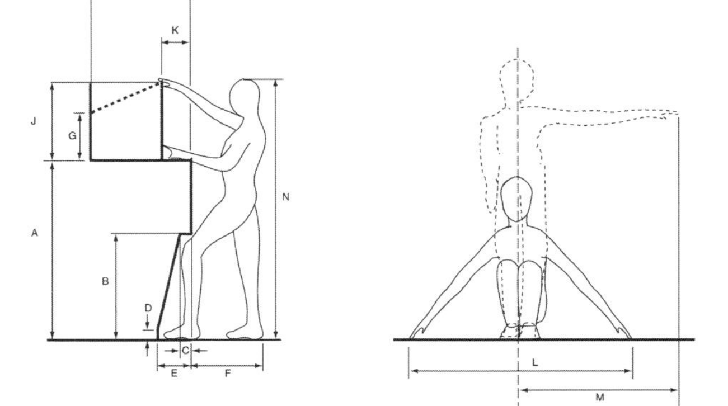

Code	Dimension	* Percentile men (in mm)			* Percentile women (in mm)			Comments
		5	50	95	5	50	95	
A	Work surface height (elbow)	970	1070	1190	905	995	1065	
B	Knee clearance height			560				Minimum required
C	Knee clearance depth			125				clearances
D	Toe clearance height			100				for all users
E	Toe clearance depth			250				
F	Back clearance depth			915				
G	Max reach access height	380			380			
H	Max reach access distance	915			890			
J	Upright reach access height	685			635			
K	Upright reach access distance	455			405			
L	Crouch space for side pick-up			1145				
M	Extent of max side reach[a]	1145						
N	Standing worker's height	1600	1730	1880	1505	1605	1695	

[a] For a twisted trunk, M will be less than stated.

*Percentile: the value below which falls a specified percentage (i.e. 5, 50, 95) of a large number of statistical units. In the context of this manual, it refers to bodily dimensions or to dimensions of workbenches, seats or other work facilities.

Figure 3.2   Dimensions for the adult population in the United Kingdom: (b) Clearance and access dimensions for a seated worker

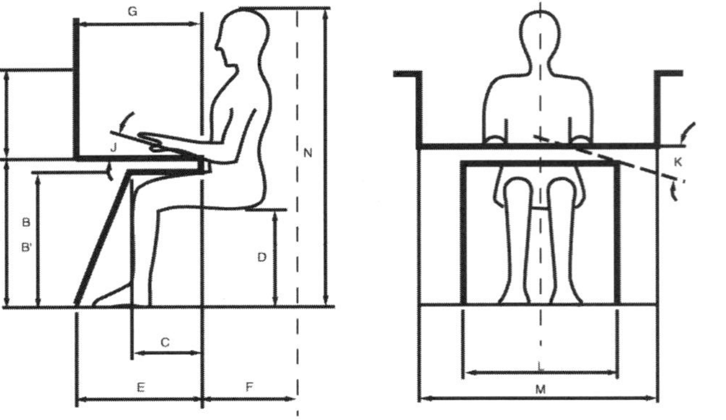

Code	Dimension	* Percentile men (in mm)			* Percentile women (in mm)			Comments
		5	50	95	5	50	95	
A	Work surface height (elbow)	660	710	760	635	685	735	
B	Knee clearance height (cross-legged)			750			725	Minimum required clearance
B¹	Knee clearance height (uncrossed)			625			600	Minimum required clearance
C	Knee clearance depth			350			350	
D	Seat height (horizontal seat)	447	483	523	424	460	495	Use 5th percentile if only using min. B
E	Foot clearance depth			650			650	
F	Clearance for getting up			635			635	
G	Grasp objects depth	610			610			Assumes slight leaning
H	Grasp objects height	380			380			Assumes slight leaning
J	Keyboard or switch panel angle		15°			15°		
K	Keyboard angle for one-handed operation		20°			20°		
L	Knee-well depth			650			650	
M	Distance apart of side panels for max. reach to grasp to height H	1650			1500			
N	Max. head height			1450				

Source: Clark and Corlett, 1984.

Specific recommendations and guidelines

As well as general and generic guidelines about designing workspaces, particular recommendations may be available for particular workplaces or types of work, or equipment or tasks. See figure 3.3 for an example which gives variations from elbow height for bench heights for different types of work.

Figure 3.3   Indication of changes in workplace height
for different types of work
Based on heights of males and females in the United States

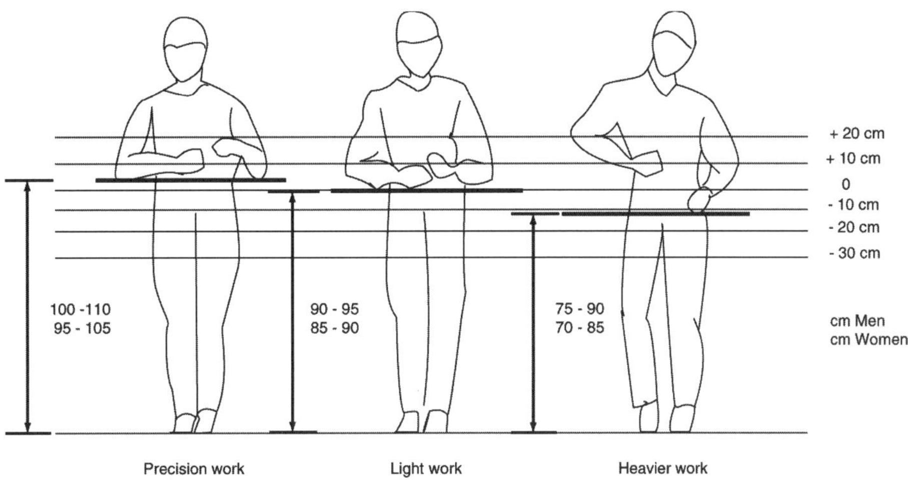

Source: Grandjean, 1979.

Key points for the design and
layout of workplaces

To completely plan for, design and implement a new workplace is a considerable effort requiring consideration of many factors. The best solutions will differ according to the type of work and workstation – office, driving, assembly, control room, etc. Therefore, only a general list of principles and guidelines to incorporate in such an effort are provided here.

The most important principles include participation, flexibility, adjustability and suitability for the widest possible range of likely users.

Following these principles, recourse can be made to specific guidelines for workplace layout, seating and the physical environment.

Guidelines for the arrangement of workplaces
(Adapted from Corlett, 1978)

1. The worker should be able to maintain an upright and forward-facing posture during work.

2. Where vision is a requirement of the task, the necessary work points must be adequately visible with the head and trunk upright or with just the head inclined slightly forward.

3. All work activities should permit the worker to adopt several different, but equally healthy and safe, postures without reducing capability to do the work.

4. Work should be arranged so that it may be done, at the worker's choice, in either a seated or standing position. When seated, the worker should be able to use the backrest of the chair at will, without necessitating a change of movements.

5. The weight of the body, when standing, should be carried equally on both feet, and foot pedals designed accordingly.

6. Work activities should be performed with the joints at about the mid-point of their range of movement. This applies particularly to the head, trunk and upper limbs.

7. Where muscular force has to be exerted, it should be by the largest appropriate muscle groups available and in a direction co-linear with the limbs concerned.

8. Work should not be performed consistently at or above the level of the heart: even the occasional performance where force is exerted above heart level should be avoided. Where light manual work must be performed above heart level, rests for the upper arms are a requirement.

9. Where a force has to be exerted repeatedly, it should be possible to exert it with either of the arms, or either of the legs, without adjustment to the equipment.

10. Rest pauses should allow for all loads experienced at work, including environmental and information loads, and the time interval between successive rest periods.

Guidelines for work seating
(Corlett, 1995, and Pheasant, 1988)

1. Seats should provide stable bodily support, allowing the feet to support less than a third of body weight, in a posture which is appropriate to the activity, physiologically satisfactory and comfortable over a period of time.

2. Seats should allow changes in posture with and without adjustment of the seat.

3. Seats should be adjustable (horizontally as well as vertically if possible).

4. Seat width should cope with largest user hip widths.

5. Seat depth (length) should cope with shortest user thigh length.

6.   Seat height should be adjustable. If it must be fixed, then ensure that, with a footrest, users with shortest lower leg length can rest feet horizontally and that users with longest lower leg length have space to place legs under worktop and feet on floor.

7.   An adjustable backrest must provide support for the lower back.

8.   All adjustments must be quick and easy to make (gas lift seats are good).

9.   Seat surfaces should be lightly padded, covered with non-slip material and with a "waterfall" edge at front.

10.   Seats should swivel and be easy to move (unless large horizontal forces are to be applied).

Guidelines for the physical work environment

The three principal factors in a typical work environment are lighting, climate and noise. Environmental problems from vibration, radiation, chemical and dust sources are not considered here since they require specialist advice; so do advanced engineering aspects of all environment factors.

The simplest form of environment survey of a workplace or room is to map the area and take relevant readings. Mapping the environmental problems provides a clear picture of where and what to control. First, a plan view of the area is drawn, marking on it the positions of machines, windows, doors, light and heat (cold) sources. Then, illuminance, noise level in dB(A), and climate (temperature, humidity, air movement) measures are taken at regular intervals over a standard working day or shift. Positions for measurement will be determined by overlaying a regular grid pattern on the room plan, with readings taken at about two to three metre intervals. Recordings are also made at all key work task sites. From this mapping, an environmental profile of the whole area can be produced, with equal value contours drawn where appropriate, and point values identified at work tasks sites.

Lighting (CIBS, 1984)

1(a)   Illumination falling onto the task must be according to good practice as in the code for interior lighting.

Typical illuminance values should be:

Waiting rooms, storerooms	200-300 lux
General engineering	300-500 lux
General offices	500 lux
Drawing boards	750 lux
Inspection, very fine assembly	1,250 + lux

1(b)   Values may be increased or lowered slightly according to the age of users, criticality of task/safety, length of task time, etc.

1(c)   Consideration must be given to avoiding all glare – direct or reflected – from windows, lights or polished surfaces.

1(d)   Lights must not have perceptible flicker.

1(e)   Colour and object contrasts must be assessed for comfort and for performance in inspection tasks.

Climate

1(f)   Daylight sources must be provided where possible.

2(a)   Climate factors at work are ambient temperature, radiant temperature, air movement and humidity. These must be set in conjunction with clothing worn and activity levels.

2(b)   Recommendations for comfort include:

Ambient temperature:	seated, mental work	20°C-21°C
	seated, light work	18°C-19°C
	standing, heavy work	15°-17°C
Radiant temperature (acceptable range):		16°C-20°C
Humidity:	45-65% at 18.5°C is acceptable	
Air movement:	comfortable: 0.11 - 0.15 m/s	
	("draughty": 0.5 m/s)	

2(c)   If work conditions are very warm or cold, then thermal stress and work disability will be problems. Some relief can be found through appropriate clothing and frequent rest breaks, but specialist advice must be sought.

Noise

3(a)   Limits on exposure to noise are a function of the sound intensity and frequency, duration of exposure, and continuous or intermittent nature of the signal.

3(b)   Intensity is measured on a logarithmic scale, typically in dB(A)s (decibels weighted for response of the human ear); an increase of 3dB(A) then reflects a doubling in sound intensity.

3(c)   Legal limits differ in different parts of the world. Swedish Standard 590111, provides the following guidelines:

>5h exposure/day – 85dB(A) max
2-5h exposure/day – 90dB(A) max
1-2h exposure/day – 95dB(A) max
<20m. exposure/day – 105dB(A) max

3(d)   For performance interference, limits should be much lower. For example, in order not to interfere with speech communications, background noise should always be 10dB(A) to 20dB(A) below speech intensity; for normal speech at two metres' distance, this means a maximum of about 60dB(A) and, at six metres' distance, a maximum of 50dB(A).

Principles and approaches to workplace evaluation and design

The keys to successful workplace evaluation and redesign are participation of all involved in the generation of solutions, and flexibility and adjustability in the solutions themselves. In general, the less rigid the work, work methods and work equipment, the more adjustability is possible. Providing people with the potential to change tasks, work methods, position of work, posture, arrangements of work equipment and furniture, and environmental variables will improve performance, health, satisfaction, comfort and attitudes.

Broadly, we will be looking to redesign workplaces according to one or more of four basic principles.

Design for extreme cases

Here we would design for a "worst" or near-worst case, in the knowledge that the solution will be acceptable for everyone else. For instance, an access hatch for maintenance will be of a size to let the largest user through (wearing protective clothing if relevant); a vital control would be placed within easy reach distance of the person with the least arm reach (not necessarily the smallest person); noise exposure levels should reflect the needs of the most hearing-sensitive workers.

Usually design for extremes is actually for a near-extreme value, say 97.5 per cent of users. Doorway heights of 240 cm to fit the very occasional basketball player would be too costly and have too many technical implications for building construction to be worthwhile, given the very small number of people of stature over 190 cm.

Design for adjustable range

It may be that we cannot find one design or dimension to suit the vast majority of users. Then we would look to provide adjustment for the critical factor. Examples are adjustable seat heights, lighting at the task (rather than the whole room) which can be adjusted in intensity and direction for individual preference, modular work equipment (e.g. at supermarket checkouts) rearrangeable by the individual workers, tiltable car seats, etc.

As above, adjustable range including absolute best values does not usually provide for 100 per cent of potential users, for cost-benefit and other reasons. Seat height adjustment range, for instance, may allow optimum positions for 95 per cent (2.5 percentile to 97.5 percentile) of users. If the acceptable sitting height for each of a large group of people was measured and plotted, we would find most measures towards the centre and the results tailing off to a very few at the high and low seat height levels. If we now take all (100 per cent) of our sample, and cut off the lowest 2.5 per cent as well as the highest 2.5 per cent, we are left with 95 per cent of the sample. The seat height at the point where the 2.5 per cent was cut off is referred to as the 2.5 percentile and that, at the 97.5 per cent point (2.5 per cent from the high seat height end) as the 97.5 percentile.

An extension of the adjustability principle is where a variable is changed according to the needs of most of a working group; e.g. climate control in an office or factory floor.

Design for restricted or tolerance range

In many situations, individuals may have a fairly wide range of tolerance for a specific design dimension or else designing for adjustability may be prohibitively expensive. In such circumstances, it may be appropriate to design to a "common area of fit"; that is, the dimension or layout used will not necessarily be optimum for everyone, but will be satisfactory for all or most users. Examples are a simple workbench height for standing work or the ambient light levels in a room. Such a compromise may be arrived at due to different criteria for different people: plug sockets in public buildings, for instance, should be too high for very young children to reach, within a range accessible to wheelchair users, and not so low or high that bending or stretching to reach is awkward for adults.

Designing for an average value

As a last resort, and as a particular case of designing for a tolerance zone, we may wish to take an average requirement in ergonomics design. One example may be the height of displays to be viewed by standing operators; even here though, it may be that it would be better to design higher or lower than the average after taking into account critical use and user factors.

Testing and evaluating workspaces

It is now time to turn to evaluation procedures and to give attention to situations where a workplace is being redesigned or evaluated for its effectiveness. In such a situation, people with experience of the situation are present and can be brought in to the redesign process, where their expertise can contribute to improving the design. Where this is not the case and the new workplace has no antecedents, there are useful procedures for testing its effectiveness before mistakes are converted to metal.

One simple method of evaluating workspaces is the "walk through". Given that a drawing or model of the layout of the workplace exists, with its equipment, supply and delivery routes decided, then the designers, industrial engineers and operators can simulate the work process.

Using the task synthesis and information on the proposed work organization, each step in the job – together with emergency, stop and start, and maintenance procedures – is simulated as if it was being done. Any questions which arise are noted – no judgement or decision is made initially – for subsequent discussion. If each person, in turn, does this while being watched and listened to by the others, a number of questions are gathered from which useful improvements can be developed.

Fitting trials

Fitting trials are a highly desirable procedure for any new or redesigned workplace. They check the goodness of fit between the design and its users and, particularly where anthropometric data are insufficient, ensure both a good match and define the range of adjustments which will be needed. It is a logical procedure which should, where possible, be carried out in an area accessible to other operators so that their comments can be obtained.

The procedure is as follows:

1.   Construct a mock-up of the workplace. Provide adjustability of major design features.

2.   Select a sample of subjects to reflect eventual user population, preferably actual or intended users. If the design concerned is only uni-dimensional, you can select a small homogenous sample. For example, you want only to know the recommended size of a hatch for entry/exit, then take a sample of (say) six or eight 95th percentile body size males since all smaller will fit; if you are concerned with work surface height, then take 12 to 25 subjects spread across the 5th, 50th and 95th percentile groups.

3.   Record anthropometric data of sample.

4.   Decide order of adjustment of workplace features.

5.   Be consistent in your description of the tasks to be performed.

6.   For each subject, move the adjustable component at discrete intervals throughout its range, from very little/low/small to very great/high/large, and then in the other direction. Have them indicate whether the component or dimension is "too large/high, etc" or "too little/low, etc." or "satisfactory".

7.   Plot results for all subjects to find a fixed satisfactory dimension or a common range of adjustment.

8.   Repeat for all other adjustable components, with previously tested dimensions in optimum or satisfactory setting.

9.   Specify workplace and evaluate.

In figure 3.4, you will see that two male and two female subjects were used in each of the 5th, 50th and 95th percentiles for standing height. Each of the 12 subjects experienced the chair being adjusted up and down, in 25mm steps from about 600mm to 1100mm heights. Do *not* start the reversal of the direction from the point you finished the previous trial, your subject may have counted the number of trials since the last "acceptable" report, and bias the results by counting back on the second trial!

The cross shows where each person reported the height as "most comfortable," and the vertical line represents the range of "acceptable" heights. If one horizontal line could be drawn through all of these, no adjustment would be needed. In the example shown, this could not be done, but the two horizontal lines just touch all the "acceptable" reports from all subjects, and thus represent the *minimum* amount of adjustment necessary to suit 90 per cent (5th to 95th percentiles) of the population.

Figure 3.4 Seat height for a work height of 105 cm

Highest preferred height: .... cm   Adjustment range: .... cm
Lowest preferred height: .... cm
Average preferred height: .... cm

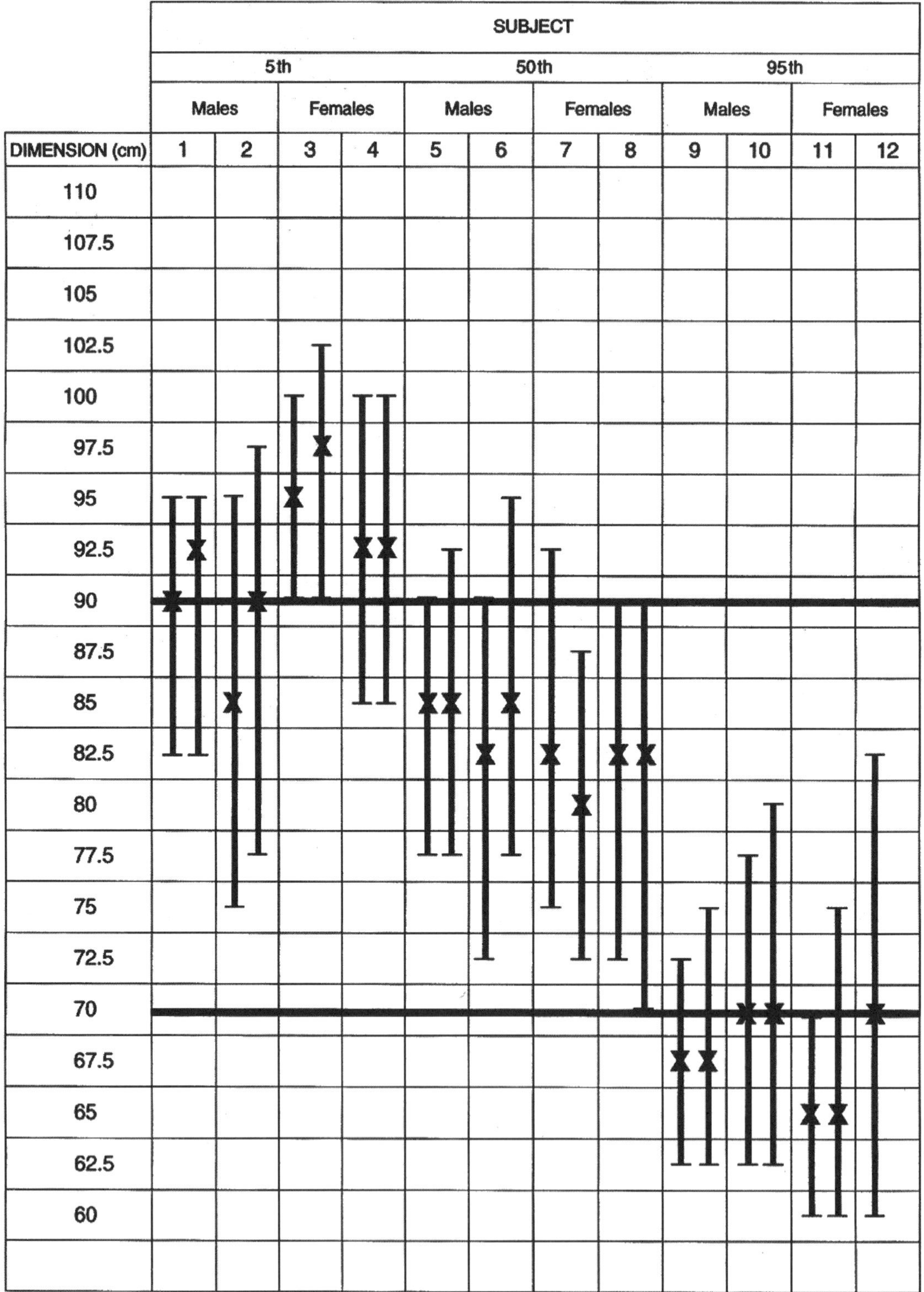

DIMENSION (cm)	SUBJECT											
	5th				50th				95th			
	Males		Females		Males		Females		Males		Females	
	1	2	3	4	5	6	7	8	9	10	11	12

Source: Pheasant, 1994.

User trials

Extending from fitting trials is the notion of user trials. In the sense that almost all evaluation and design exercises in ergonomics preferably should include user trials, then ergonomics practice always has some participative element. (Exceptions to the requirement for user trials are potentially hazardous systems, or systems where tests with users may be prohibitively expensive or unethical. Here we would use some form of simulated ergonomics testing.)

User trials will be set up to test more than single aspects of a workplace, as is the case in fitting trials; indeed, fitting trials may well be one part of a user trial. Typically a sample of actual or potential users, or of people similar to the expected users, are brought together to test a product, system or workplace. This is generally done in a systematic controlled manner, with the effects of the key variables being noted by taking some predefined appropriate measures. User trials are an integral part of an ergonomics design process, allowing reasonably valid assessment of products and workplaces *during*, not after, the design process.

A group of subjects is selected to represent the user population; the size of the group should be appropriate for the resources, complexity and desired outcomes of the investigation, but may number anywhere from five to 100. Tasks must be developed to realistically represent actual work to be done, but to be simple enough for "new users" to perform adequately. The test workplace layout may be more or less similar to a finished version; it could just be a rough model, for instance.

The subjects will be asked to perform certain tasks in a certain order. Measurements will be made by direct observation, indirect observation through use of interviews, questionnaires and rating scales, and performance or behaviour measurement. Criteria chosen may be time, accuracy and errors, convenience, ease of use, comfort, satisfaction, potential health and safety, physical fit or clearance or reach, physical workload, etc. Subjects will always be debriefed at the end of the trials.

Critically, certain dimensions or features of the workplace, or the whole workplace itself, usually will be altered within the trial. Thus the evaluations of different design features can incorporate the differences between the decisions of different people, and how an individual's view can change with changes in certain circumstances. Overall, the user trial is a relatively quick, low-cost and efficient aid to better development and prototyping of workplaces. Within the limitations pointed out, such as in cases of hazardous work systems, involving workers in the process of redesigning workplaces is highly rewarding.

Checklists

Direct observations of tasks and workplaces, subsequent analysis and evaluation of changes made can be greatly assisted by consistent use of a rational, comprehensive checklist. As with assessment of work tasks and postures (see Chapter 2), such checklists are best adapted by companies for their own use. However, companies can learn from other experiences and techniques.

Checklists can act as reminders, to be sure that all points have been covered, or as a means for investigating or evaluating a situation. An example of the former is the visual environment checklist in box 3.1.

The checklist in table 3.2 combines both acceptability assessment and profiles of the mismatch on certain dimensions. It will be evident that many of the workplace questions could have been changes which would have been desirable. In general, before asking questions about satisfaction (or otherwise) with features, it is best to think through what you will do with the answers. When investigating attitudes, satisfaction is an important dimension, as is pointed out in Chapter 5. Where physical or technical aspects are issues, although satisfaction is important, it often arises from a well-designed relationship with the equipment, etc. Hence, the use of scaled-feature checklists are advantageous.

Obviously a checklist can be combined with other techniques to provide a worksheet for evaluation and development. A simple example is shown in table 3.3. In the latter, the investigator worked with the operator to decide on the quality of each of the workplace factors studied, an example of a built-in cooperation procedure.

Box 3.1   Checklist for evaluation of any visual environment
(N.B. Some points will not be relevant for some environments)

What are the problems?
What measures are needed?
What guidelines are available?

Consider each of the following if appropriate:

Light levels
Luminance and illuminance:
    Where is the light needed?
    What variation is there?
        across the room?
        across the workplace?
    Is supplementary light needed anywhere?
        at any particular time of the day?
        at any particular time of the year?
        for particular purposes related to the immediate task?
        for purposes unrelated to the tasks, e.g. safety lighting?

Surfaces

What are the reflectances of the various surfaces?
the walls,
the ceiling,
the floor.
What are the reflectance ratios between them?

Glare

Discomfort glare:
Is it a problem?
Subjective or objective assessment needed?
Can it be relieved by modifying the environment?

Disability glare:
Is it affecting performance?
Can it be relieved by moving or shielding the lights?

Temporal aspects

Is flicker apparent:
in the visual task itself, e.g. a VDU?
in the environment, e.g. from fluorescent tubes?

Chromatic considerations

Is colour rendering of concern?
Is the colour rendering of the lights acceptable?
Is the colour appearance of the lights acceptable?
Is colour discrimination a factor for concern?
How good is the colour rendition of the present lights?
Is the supplementary light with better colour rendition needed for the task, to enhance colour discrimination?
Is colour coding present, and if so are the lights adequate for the colours to be discriminated?

Spatial considerations

Directionality:
Is there a specific need for directional lighting?
What variation is there over the room?
What variation is there over the day? How is the daylight supplemented by artificial light, and how does this vary over the day?
Are shadows a problem under working conditions?

Highlighting:
Are there any features that need special consideration to increase their conspicuousness?
Are there any special-purpose needs, such as safety lighting with a back-up power supply?

Users

Are there particular user groups with specific requirements?

Non-visual considerations of the visual environment

How can the environment be maintained under the desired conditions?
What maintenance is needed?
What non-visual effects occur, e.g. heat from luminaries?

Source: Howarth, 1995.

Table 3.2   A workplace and work environment questionnaire

WORKPLACE LAYOUT

This part of the questionnaire requires you to rate your workplace on a five-point scale. There are a number of statements about your workplace. Complete each by circling the number which corresponds to whether you agree or disagree with the statement about that particular aspect of your workplace.

1 — strongly agree
2 — agree
3 — uncertain
4 — disagree
5 — strongly disagree

For example, if a statement said "The chair material is comfortable" and you disagree, although not strongly, then you should answer as follows:

The chair material is comfortable   1   2   3   (4)   5

1.   Using the following scale:

1 — strongly agree
2 — agree
3 — uncertain
4 — disagree
5 — strongly disagree

How do you view the following aspects of your workplace?
(Please circle the appropriate number)

	1	2	3	4	5
The leg room underneath the desk is adequate	1	2	3	4	5
The chair adjustability is good	1	2	3	4	5
The backrest is good	1	2	3	4	5
The footrest is suitable	1	2	3	4	5
The arm support is comfortable (where relevant)	1	2	3	4	5
The hand support is comfortable (where relevant)	1	2	3	4	5
Overall, the chair is comfortable	1	2	3	4	5
The keys on the keyboard are easy to press	1	2	3	4	5
The desk is easy to use	1	2	3	4	5
It is easy to correct errors	1	2	3	4	5
Errors, when made, are visible	1	2	3	4	5
The desk as a whole is well-designed	1	2	3	4	5

Table 3.2   (cont.)

2. Now please answer the following more detailed questions about your workplace.

How do you view the following aspects of your workplace?

Height of the desk (tick one)	Too high	☐ (1)
	About right	☐ (2)
	Too low	☐ (3)

Height of the chair (if not adjustable) (tick one)	Too high	☐ (1)
	About right	☐ (2)
	Too low	☐ (3)

Height of the keyboard on the desk (tick one)	Too high	☐ (1)
	About right	☐ (2)
	Too low	☐ (3)

Height of the work holder (tick one)	Too high	☐ (1)
	About right	☐ (2)
	Too low	☐ (3)

Height of the input surface for work (tick one)	Too high	☐ (1)
	About right	☐ (2)
	Too low	☐ (3)

Position of the output tray/surface (tick one)	Too far away	☐ (1)
	About right	☐ (2)
	Too near	☐ (3)

Table 3.2   (cont.)

WORK ENVIRONMENT

Now please answer the following questions concerning your workplace environment:
in **SUMMER** and **WINTER**.

		Summer	Winter
Direct lighting at the desk is	much too bright	☐ (1)	☐
(tick one)	too bright	☐ (2)	☐
Additional comments	comfortable	☐ (3)	☐
....	too dark	☐ (4)	☐
....	much too dark	☐ (5)	☐
....			
The general lighting in the room is	much too bright	☐ (1)	☐
(tick one)	too bright	☐ (2)	☐
Additional comments	comfortable	☐ (3)	☐
....	too dark	☐ (4)	☐
....	much too dark	☐ (5)	☐
....			
Room temperature is	much too warm	☐ (1)	☐
(tick one)	too warm	☐ (2)	☐
Additional comments	comfortable	☐ (3)	☐
....	too cool	☐ (4)	☐
....	much too cool	☐ (5)	☐
....			
The noise level at the desk is	much too loud	☐ (1)	☐
(tick one)	too loud	☐ (2)	☐
Additional comments	comfortable	☐ (3)	☐
....	too quiet	☐ (4)	☐
....	much too quiet	☐ (5)	☐
....			
Ventilation at the desk is..................................	much too great	☐ (1)	☐
(tick one)	too great	☐ (2)	☐
Additional comments	comfortable	☐ (3)	☐
....	too little	☐ (4)	☐
....	much too little	☐ (5)	☐
....			

Source: This checklist has been provided by J. Wilson.

Table 3.3   Part of a checklist used on an ergonomics training course for improving working conditions and productivity in small enterprises

	Do you propose action?			Remarks
	No	**Yes**	**Priority**	
1. Clear everything out of the work area which is not in frequent use				
2. Provide convenient storage racks for tools, raw materials, parts and products				
3. Use specially designed pallets to hold and move raw materials, semi-finished goods and products				
4. Put stores, racks, workbenches, etc., on wheels for easy handling				
5. Use carts, movable racks, cranes, conveyors or other mechanical aids when moving heavy loads				
6. Put switches, tools, controls and materials within easy reach of workers				
7. Use lifts, levers, or other mechanical measures to reduce the effort required by the worker				
8. Provide a stable surface at each workstation				
9. Use jigs, clamps, vices or other fixtures to hold items while work is done				
10. Adjust the height of equipment, controls or work surfaces to avoid bending postures or high hand positions				
11. Change working methods so that the workers can alternate standing and sitting while at work				
12. Provide chairs or benches of correct height with a sturdy backrest				
Source: Thurman et al., 1988a.				

Case-study 3A illustrates an example of the use of checklists and questionnaires, and shows that, when well-designed and used, they can be reliable in the results which are revealed. Pilot trials will often reveal where the flaws lie in checklists.

CASE-STUDY 3A

Work conditions and health issues in VDT use

The ABC company was concerned to assess health and safety considerations associated with use of visual display terminals (VDTs) by groups of its employees. The ergonomists involved in the study set out with agreement of the company to:

- discuss the study openly with employees at the outset

- share results with relevant trade unions

- investigate work organization and workplace layout factors, as well as health reports

Typical of such investigations, the study had two components: a confidential questionnaire survey of operators administered at three intervals over about two years, and an ergonomics survey of every VDT workstation in the company. The questionnaire covered employee perceptions of work organization (e.g. workload and pacing), conditions (environment, etc.), job content, supervision, social contact, technology used, general career considerations, health status, and health and psychological complaints. Importantly, the questionnaire was examined question by question by the union and the company before administration, and each question's importance justified. A checklist guided the workplace survey, looking at physical workstation dimensions, worker posture, work environment measures and features' presence or absence.

Good compatibility was found between the two study methods; both identified problems with screen glare and reflections, workstation and seating design (dimensions and adjustability), noise, and (by the operators only) temperature and humidity. Despite this apparent duplication, the dual methodology is necessary to give some cross-validation and to breed feelings of both participation and scientific status among the workforce and management.

Recommendations were made concerning all inadequate factors found, and also about training and support for operators and supervisors covering VDT work, equipment and workstations, and ergonomics improvements. Based on these recommendations, the company started implementing changes, with positive outcomes for employee perceptions and attitudes.

Source: Carayon et al., 1991.

Redesign and participative procedures

The physical workspace and environment offer the ideal focus for participative work redesign practices, with actual users involved in the diagnosis of existing problems, collection of data, generation and testing of solutions. Many ergonomics processes are to some extent participative (user trials or workplace attitude questionnaires), but best results will be achieved when all or most staff are involved continuously (or at least regularly) in analysis, development and evaluation exercises.

A participatory process is not, however, an easy option. Half-hearted or cynical adoption and application will reap a harvest of suspicion, reluctance and disrepute. Possibly the key success factors are the nature of the facilitator, the ability to give the process enough time, the degree of competence and motivation of the participants, and the "visibility" of the problem.

Workers' involvement is central to the redesign of existing workplaces because better solutions will result more quickly with the involvement of those directly affected and those who know most about workplaces and related problems. Moreover, information and consultation will foster the commitment and dedication of the workforce (see case-studies 3B and 3C).

Where companies are small some of the foregoing may appear impractical, if only because the company's resources are not large enough to introduce some of the suggested procedures. Yet there is still much that can be done with some of the simpler procedures by company management. Studies by the ILO are available which illustrate cases of low-cost changes to improve performance and working conditions (Kogi et al., 1988). The ILO has also produced a series of training manuals for a very straightforward programme of improvements to workplaces and productivity which includes a checklist approach to analysis for many of the problem areas given earlier (Thurman et al., 1988a), or a series of ergonomic checkpoints for workplaces (ILO, 1996).

CASE-STUDY 3B

Ergonomic improvements through workers' involvement

Plant X is an incinerator plant which takes refuse from across a major city in the United Kingdom and uses it to produce power and heating for many of the city centre's public buildings and private houses. Movement and loading of the refuse is carried out via giant grab cranes. These are controlled by a very few of the site personnel, sitting some 20 metres above the refuse and hoppers in a small control room. Following discussions between the operating company and the relevant union health and safety officer, some outside ergonomics consultants were employed to investigate and recommend (preferably low cost!) solutions. The investigators sought agreement that the study would be, as far as possible, a participative exercise involving the crane operators. To do this, they had to overcome problems of availability (there are only four to five crane operators, of whom one was working, one sleeping and one on holiday at any time), scepticism and low educational attainments among the workforce.

Early observation of the work indicated that the problems for the operators were to do with viewing the task, physical demands and posture, and general environmental input. In order to promote two-way participation and gain operator confidence, the principal investigator trained the operators to perform a limited set of work tasks. This work, and observation of a number of tasks, enabled a task analysis to be produced. Various recording and analysis techniques were employed; these included a basic ergonomics workplace checklist and related questionnaire, a body part discomfort survey (see figure 2.4) posture recording, photography and video analysis.

Through all these methods, the major problems identified were:

- the position and angle of the viewing window, and the position of the primary task (picking up loads) causes a very poor work posture, leaning forward out of the seat;

- this is compounded by the design and position of crane controls, including a "dead man's handle", which are placing considerable static posture stresses on the operators;

- poor lighting, lack of daylight and spectral reflections on the viewing windows give an ineffective and dissatisfying work environment;

- the position of the entrance to the control booth led to considerable atmospheric dust build-up.

On the basis of data collected, a number of participative redesign sessions were initiated. In one, and despite severe doubts as to its appropriateness, three of the operators and two investigators had an extended design decision session. A number of brainstorming and creative techniques were used, including drawing and critiquing the existing and possible new workplaces. Several similar workplaces at other companies were visited by the investigator and at least one operator. A multi-configurable work seat and crane control set was built and, in development and fitting trial sessions with individual workers, alternative improved control layouts and workplaces were built and assessed and a preferred layout defined. Also, a very fast and cheap procedure was implemented to reconfigure the physical environment, especially the visual environment. Moveable screens, blinds and floor coverings allowed the crew to develop recommendations that minimized spectral reflections off the viewing glass and improved ambient illumination and the appearance of the visual environment.

All recommendations were to be used as the basis of a redesign process within the company. The key to the successful generation of feasible solutions was the full and enthusiastic participation of the operators. This was better enabled because they were such a small well-knit group and because the principal investigator became almost a trainee operator.

Source: Moreland and Wilson, 1992.

CASE-STUDY 3C

A packaging workstation: Lessons learned the hard way

It is impossible to get every ergonomics intervention programme to work perfectly from the beginning each time. Often it is a case of welcoming the improvements that are realized, but learning for next time from things that go wrong.

Preliminary investigation showed problems with absenteeism, physical exertion and job satisfaction among female packers of tobacco cartons at the Dutch foodstuffs manufacturer. A multidisciplinary project team was set up comprising relevant managers and foremen, as well as health and safety and ergonomics professionals. Following a problem definition phase, a "situation analysis" was conducted, including task analysis, checklist evaluation of workstations, evaluation of postures and motions against two checklists, evaluation of the lifting work against NIOSH (United States National Institute for Occupational Safety and Health) norms, and evaluation of job characteristics looking at the challenge and interest in the work. Mock-ups of a proposed workstation redesign were evaluated by foremen and packers. A trial packaging workstation was implemented, which had an improved layout and was based on a change to a standing operator.

Thus the investigators did many appropriate things: setting up a project team (although not including workers in this), examining psycho-social as well as physical factors, and testing out the proposed modifications. However, extension from the trial workstation to wider implementation has not occurred despite enthusiasm in the workforce. The investigators ascribe this particularly to the fact that:

■	there should have been closer cooperation between the project team and the construction department responsible for building and installing the new workstation; and

■	workers, both those directly and indirectly affected, should have been involved more from the outset.

Both these lessons can be taken as general rules for any workplace redesign project.

Source: Mossink, 1990.

Design decision groups

Many companies use a variety of techniques to enable workers to participate successfully and develop new workplace layouts for themselves. Typically, such techniques aim to aid creative thinking and problem-solving, and breed a dynamic, relaxed, directive, focused process without being autocratic.

One typical technique of this sort is **design decision groups**. The description which follows concerns a group expecting new equipment (see Wilson, 1991c).

Process

Groups of six to eight people working with relevant current technology and workplaces meet with a couple of facilitators for about three to four hours three times over two weeks or so. The whole process runs as follows:

Session 1

Introduction This provides a general overview of design objectives.

World map Participants are asked to make an inventory of all the equipment required at their workplace.

Round robin questionnaire

Participants are presented with a series of very simple open-ended statements about the work, etc., which are passed around to everybody, with each participant attempting to complete the sentence with a different ending to that chosen by the others.

Individual silent drawing exercise

Each participant is required to produce sketches of both plan and perspective views of their ideal workplace incorporating the new equipment. At this stage, only general ideas need to be produced. It is beneficial to discourage communication at this point, in the hope that a variety of different ideas and designs will be forthcoming. The confidence necessary to produce such drawings may be promoted, and further useful information gained as a by-product, by the participants first drawing their own workplaces, highlighting good and bad points.

Individual presentation and discussion

Each participant is required to present a final sketch to the rest of the group, to explain the reasoning behind the design and to discuss this. After all the sketches have been discussed, the facilitator should then lead a general discussion, bringing up any points that may not have been previously considered.

Session 2

Drawing in pairs

In this session, the group is divided into pairs who are required to produce a larger sketch, with more emphasis on detail, such as the positioning of equipment and seating arrangements.

Presentation and discussion

Each drawing is presented to the rest of the group by its authors. Discussion at this point is directed towards choosing the best proposal, but no firm decisions need to be reached yet.

Simple layout modelling using cardboard models

This is conducted in a setting similar to that of the actual workstation. Participants are given strong boxes or cardboard to model their work equipment, a number of mock-up work counter modules, and real chairs and stools.

Discussion

Geared towards eliminating inferior designs and forming a consensus of opinion as to the optimal layout.

Session 3

Building a complete mock-up of the system

This should continue the work of the last session, incorporating further details such as workplace height and depth, and placement of ancillary equipment.

Discussion and modifications

Considering the implications of additional aspects, making modifications, and deciding on best options.

Simulation trial and assessment

A detailed assessment of the workstation design, with participants taking turns to act as operators (and customers). Fitting trials also take place.

References

R.W. Bailey: *Human performance engineering* (Englewood Cliffs, Prentice Hall, 1989).

S. Caplan: "Using focus group methodology for ergonomic design", in *Ergonomics*, Vol. 33, 1990, pp. 527-533.

P. Carayon et al.: *Physical working conditions and health status at the ABC Company*, report prepared by the Department of Industrial Engineering (Madison, University of Wisconsin-Madison, 1991).

Charted Institute of Building Services: *Code for interior lighting* (London, 1984).

T.S. Clark and E.N. Corlett: *The ergonomics of workspaces and machines* (London, Taylor & Francis, 1984).

E.N. Corlett: "The human body at work: New principles for designing workspaces and methods", in *Management Services*, Vol. 22, No. 5, 1978, pp. 20-52.

—: "The evaluation of industrial seating", in J.R. Wilson and E.N. Corlett (eds.): *Evaluation of human work*, 2nd edition (London, Taylor & Francis, 1995).

C.G. Drury and B.G. Coury: "A methodology for chair evaluations", in *Applied Ergonomics*, Vol. 13, 1982, pp. 195-202.

Eastman Kodak Co.: *Ergonomic design for people at work: Volume 1* (Rochester, 1983).

Finnish Institute of Occupational Health: *Ergonomic workplace analysis* (Helsinki, 1990).

E. Grandjean: *Fitting the task to the man* (London, Taylor & Francis, 1979).

P.A. Howarth: "Assessment of the visual environment", in J.R. Wilson and E.N. Corlett (eds.): *Evaluation of human work*, 2nd edition (London, Taylor & Francis, 1995).

ILO: *Ergonomic checkpoints: Practical and easy-to-implement solutions for improving safety, health and working conditions* (Geneva, ILO, 1996).

G. Kanawaty: *Introduction to work study* (Geneva, ILO, fourth revised edition, 1992).

K. Kogi: "Participatory training for low-cost improvements in small enterprises in developing countries", in K. Noro and A. Imada (eds.) *Participatory ergonomics* (London, Taylor & Francis, 1991).

—, W.O. Phoon and J.E. Thurman: *Low-cost ways of improving working conditions: 100 examples from Asia* (Geneva, ILO, 1988).

A. Moreland and J.R. Wilson: "Energy efficient but not ergonomic: A study of an incinerator plant workplace", in E.J. Lovesy (ed.): *Contemporary ergonomics* (London, Taylor & Francis, 1992).

J.C.M. Mossink: "Case history: Design of a packaging workstation", in *Ergonomics*, Vol. 33, 1990, pp. 399-406.

M. Nagamachi: "Application of participatory ergonomics through quality circle activity", in K. Noro and A. Imada (eds.): *Participatory ergonomics* (London, Taylor & Francis, 1991).

S.T. Pheasant: *Bodyspace* (London, Taylor & Francis, 1988).

—: "Anthropometry and the design of workspaces", in J.R. Wilson and E.N. Corlett (eds.): *Evaluation of human work*, 2nd edition (London, Taylor & Francis, 1995).

J.D. Ramsey: *Ergonomics support of consumer products safety*, paper presented to the American Industrial Hygiene Association's Annual Conference, Fairfax, Virginia, May 1978.

M.S. Sanders and E.J. McCormick: *Human factors in engineering design*, 6th edition (New York, McGraw-Hill, 1987).

J.E. Thurman, A.E. Louzine and K. Kogi [1988a]: *Higher productivity and a better place to work: Action manual* (Geneva, ILO, 1988, and second impression with corrections, 1994).

— [1988b]: *Higher productivity and a better place to work: Trainers' manual* (Geneva, ILO, 1988).

H.P. Van Cott et al.: *A standard ergonomics reference data system: The concept and its assessment*, Report No. 77-1403 (Washington, National Bureau of Standards, June 1978).

J.R. Wilson: *Work redesign at a chemical plant*, unpublished case study (Nottingham, Department of Manufacturing Engineering and Operations Management, University of Nottingham, 1989).

— [1991a]: "Participation: A foundation and a framework for ergonomics?", in *Journal of Occupational Psychology*, Vol. 64, 1991, pp. 67-80.

— [1991b]: "Extending participative processes in ergonomics", in *Proceedings of International Conference on Participatory Approaches to Improving Workplace Health*, Ann Arbor, June 1991.

— [1991c]: "Design decision groups: A participative process for developing workplaces", in K. Noro and A. Imada (eds.): *Participatory ergonomics* (London, Taylor & Francis, 1991).

— and E.N. Corlett: "Participatory ergonomics", in J.R. Wilson and E.N. Corlett (eds.): *Evaluation of human work*, 2nd edition (London, Taylor & Francis, 1995).

— and S.M. Grey: "Perceived characteristics of work environment", in O. Brown and H. Hendrick (eds.): *Human factors in organizational design and management — II* (Amsterdam, North-Holland, 1986).

LAYOUT OF EQUIPMENT AND PRODUCTION FLOW

4

Felix Schmid

Production flow

So far, we have considered how to arrange work and workplaces so that people may have the best opportunities to achieve the purposes of their work. But an enterprise is a complex intermingling of people, technology, organization structures and financial operations.

It is, therefore, necessary to look at how organization and technology can be interlinked with human activities to achieve effective performance. To do this, we must look at the very basis of a manufacturing operation; the way in which to arrange production flows in a business. However, before such basic decisions should be made, it is critically important to identify where to draw the line between operations that must be effected within a company and those that can be done or should be done elsewhere. At first glance, many companies would appear to be happy with their existing supplies policy, but probing deeper many find that much of the work which caused problems due to overloaded machines or lack of skills, etc., could have been bought in from external suppliers. In many cases, external suppliers could well have been cheaper too because of economies of scale or their experience with the technology! We will, therefore, start this section by describing the "make versus buy" decision.

The make versus buy decision

The decision whether to make or buy assemblies, subassemblies and components is obvious for some components, as the knowledge and capability required to produce some products will be well beyond the current scope of the organization. In other circumstances, it may be preferable to buy in certain components in order to produce a more standardized product. Such components could include switches, indicator lights, computer keyboards, nuts, bolts, and hydraulic and electronic components. Quite often, the suppliers of such items will have designed them specifically to meet the requirements of a wide international market, and thus offer a more general, and therefore flexible, product than might otherwise be achieved in-house.

Where the situation is less clear-cut, the decision of make versus buy is often made based upon the difference in cost between making and buying the component. However, in the prevailing industrial climate, companies are finding that their customers are becoming less exclusively concerned with cost, preferring to base their decisions on other aspects of the product and service, e.g. whole life cost or reliability. The decision whether to make a component in-house or to buy from a supplier should not be taken lightly, as it can seriously affect both the quality of the product and the performance of the manufacturing organization. By making a product in-house, the company retains control over the quality and delivery aspects of a product, so that failings in these areas can be dealt with quickly and appropriately. Problems can be resolved rapidly, and accurate forecasts are often far more forthcoming than from outside suppliers. The decision to make in-house carries with it increased inventory costs and greater loading for facilities that may be already stretched. Another factor is that specialist manufacturers will often be better able to keep up with the leading edge of production in their field.

It is important to remember that subcontracted work will not necessarily release labour or processing time unless it is immediately possible to divert these resources. Equally, the process of subcontracting work will incur other costs relating to purchasing, inspection, receiving, etc. It is obviously more attractive to fill spare capacity on the shop-floor, but this could cause problems if people or machines thereby become overloaded. Evidently, external sourcing is a very complex subject. Even today, cost considerations are used in many cases to determine policy. In a way there is nothing wrong with this simplification. The important thing to remember is that it is *only* a simplification, and should be ignored wherever logic dictates that a better option exists.

The important consideration is not to reduce expenditure at all cost, but to allow flexibility. As product mixes change frequently, so will the machine loadings and therefore lead times of products. It may be beneficial to off-load some work to a local supplier every so often, or make occasional parts in-house rather than subcontracting. In situations like this, it is important to be able to rely upon suppliers. This can only be achieved by forging closer links with the supplying companies involved, discussing quality measures, lead times and prices. Japanese firms believe in carefully selecting one supplier for an activity and using them for all related work. They consider that, while this might result in slightly higher costs for some parts, these drawbacks are outweighed by the benefits of closer ties in terms of reliability, quality and lead times of the work supplied.

Basic approaches to manufacturing

There are a number of common approaches used in the manufacture of goods, each of which has its benefits and drawbacks. These methods are described under three headings: intermittent, line and cellular manufacture (see figure 4.1). Each of these will be described in terms of its impact on products, people and productivity. The choice between these methods of manufacture depends on a number of factors, including anticipated volumes, product varieties, flexibility requirements, lead time requirements, social needs of the workforce, etc. Figure 4.2 illustrates how each type of production performs against some of these constraints, but can only give guidelines as to the suitability of each method. The choice of production method must ultimately be one of preference, both of the management and workforce.

Intermittent processes

The main characteristic of intermittent processes is the jumbled nature of the flow pattern between one stage and the next. In order to achieve economies in terms of supervision and easier allocation of work, it is usual to adopt a process clustering strategy whereby similar machines or similarly skilled operators are grouped together. In such a

Figure 4.1   Approaches to manufacturing

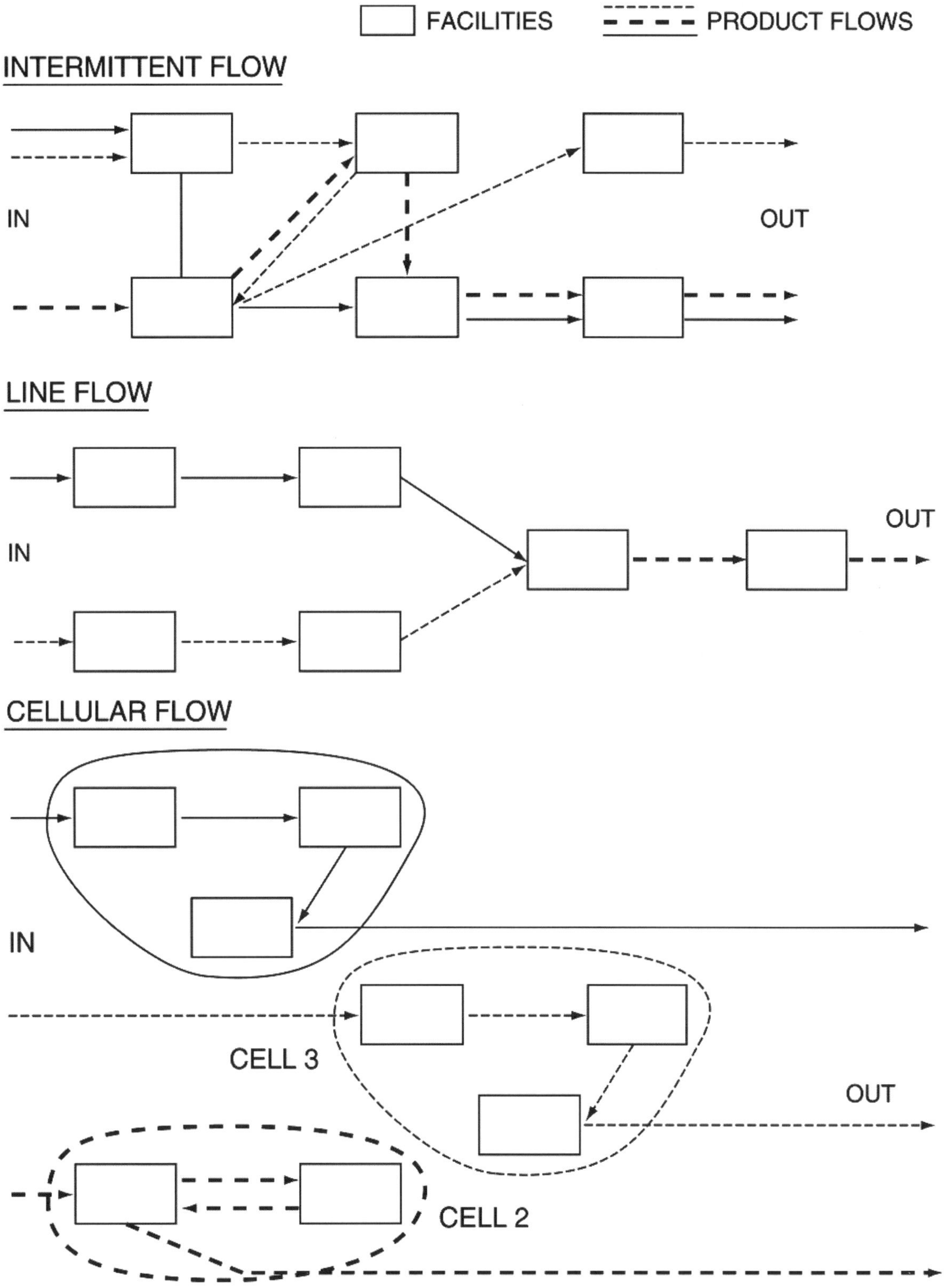

Figure 4.2   The performance of different approaches to manufacturing

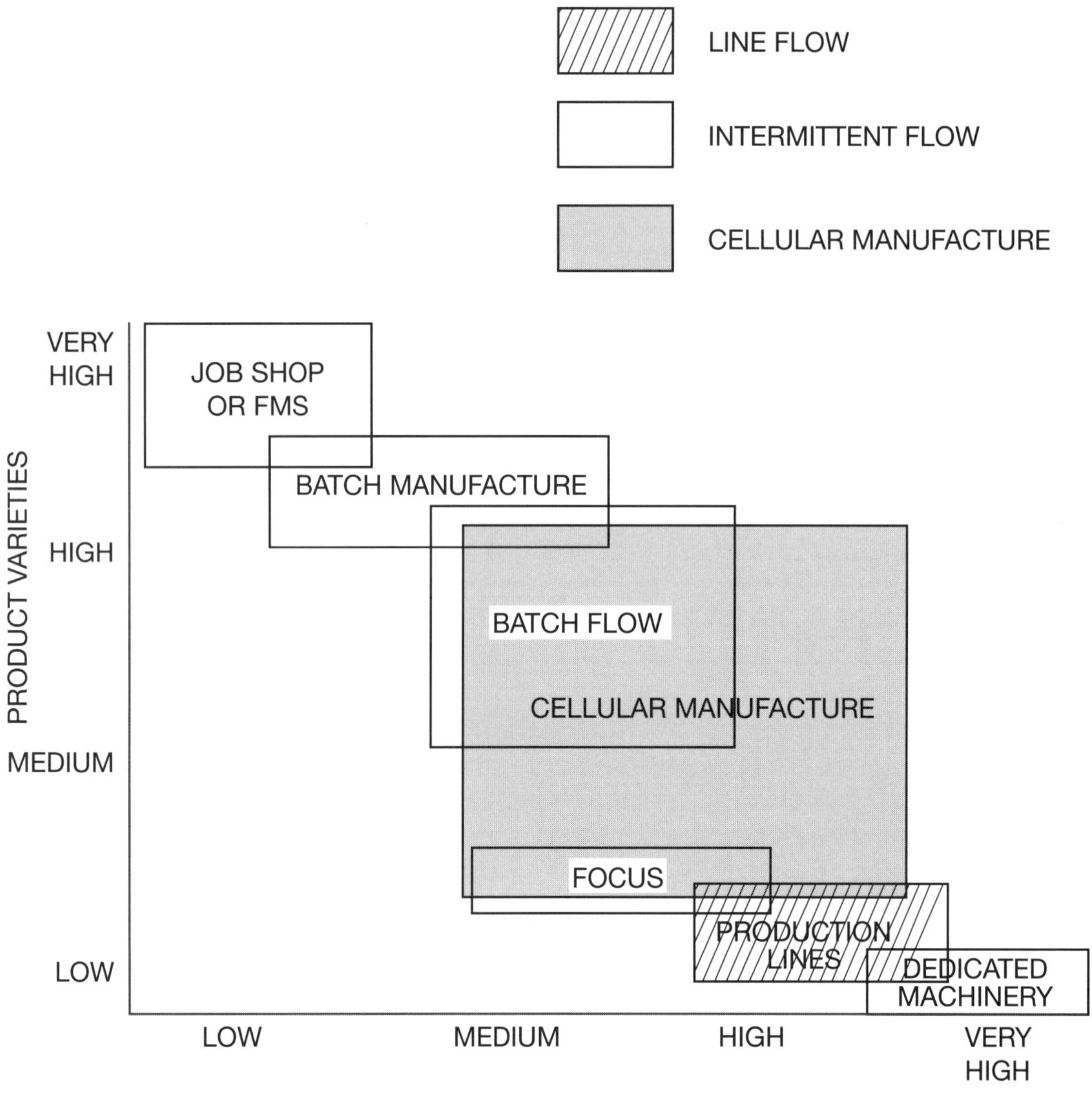

FMS: Flexible Manufacturing System

FOCUS: Focusing of production around a small number of products with low variety where output is not sufficient for production lines or where cycle times do not allow establishment of balanced flows.

situation, each workstation may be processing vastly different components, subassemblies or products within the same area. This form of manufacturing is most advantageous when a number of dissimilar products are being manufactured in small numbers, and is typified by the job-shop.

Such systems must be able to cope with frequently changing products and will therefore suffer from a need for frequent set-ups and low machine utilizations in order to achieve flexibility. This is probably the most common form of manufacturing layout, and even the most organized plant can eventually lapse into this form of working as changes are made to products and their volumes.

The main problem experienced in such systems is a high level of work-in-progress (WIP), that is, semi-finished goods or raw materials that are waiting for further processing. Thus, highly reliable scheduling of purchased parts and machine capacity becomes necessary in order to coordinate all the flows of work. This is usually done by estimating the processing times at each stage. These approximate figures can only ever give a rough idea as to delivery times, since there is a great number of possible hold-ups in the processing cycle, any of which can upset the schedule. Where large numbers of different products are produced simultaneously, it is increasingly common for companies to use computer scheduling systems such as **MRPII** (Manufacturing Resources Planning),[1] or such methodologies as **OPT** (Optimized Production Technology). While **MRPII** helps an organization to manage its resources effectively, **OPT** can often help to identify areas in which the system requires improvement.

MRPII, in particular, aims to provide a computer-based procedure for dealing with all of the planning and scheduling activities concerned with manufacturing, while also helping to keep track of cost accounting, maintenance, inventory, etc. (see figure 4.3). A key consideration when implementing **MRPII** systems is the issue of cumulative lead times: complex products with many levels in their bill of materials can result in overpessimistic lead times if buffer times are inserted indiscriminately for each level.

There are generally one or more machines within a system that are of greater importance than others and which often have a loading greater than their nominal capacity. This results in a build-up of work in front of the machine or workstation. This buffer stock consists of parts and assemblies which are often required later in the system. If this work is delayed, then the final product will be delayed also. Such machines are known as bottlenecks, and it is vitally important to find any bottlenecks in a manufacturing system so that they can be adequately handled. Bottlenecks have only limited capacity, and any capacity lost, be it through down time or processing of faulty goods, can never be recovered. Common solutions to this problem are to subcontract some of the bottleneck workload, inspect work before processing at bottlenecks and introducing shift work. Bottlenecks often exist where capacity expansion is too expensive, or the activity is not directly related to the main business of the company. Artificial bottlenecks are created when managers overestimate operation times or introduce new process plans without involving well-motivated shop-floor personnel in their decisions.

---

[1] MRP stands for Material Requirements Planning and is essentially based on the expansion of bills of materials. MRP is a method for controlling the manufacturing operation and is not a very complex tool.

MRPII stands for Manufacturing Resource Planning and is a comprehensive tool which allows the scheduling of materials and production resources using detailed information about the products to be manufactured. In general, MRPII requires a powerful computer system and the availability of bills of materials and production planning information in computer-readable form.

Figure 4.3   The elements of an MRPII system

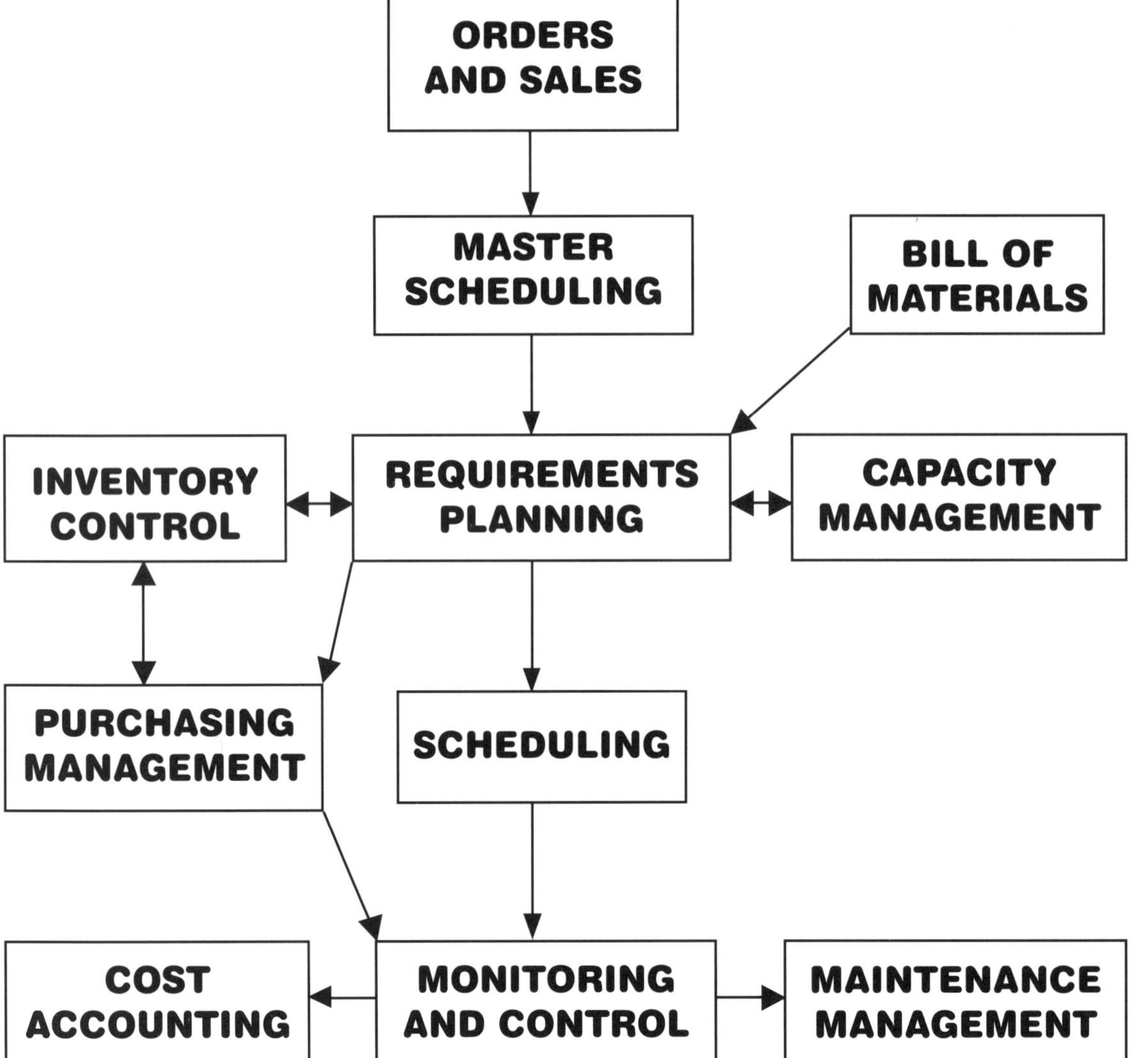

THE ELEMENTS OF AN MRPII SYSTEM

Line processes

Whereas intermittent processes are used in situations where there is a high variety of products, line processes deal with larger numbers of very similar units. The advantage of this low variety is that each unit follows approximately the same path, allowing adjacent operations to be positioned next to each other. Thus, once a product is finished at one station, it is passed to the next, maintaining a steady flow of work through the system. Set-ups are rarely necessary, with the result that very high machine utilizations are possible.

In order to ensure that work flows regularly, operations at each of the stages must be of equal duration. The output of the line in terms of units per hour is also controlled by the duration of each cycle, one unit being produced per cycle. This implies that if a high output is required of the line, then each stage has a very short duration, a situation encountered especially in the car industry.

While line manufacturing is often more efficient than intermittent processing, the type of tasks that arise are generally monotonous for the workers, especially where short cycle times are involved. This affects the morale of the workforce, and leads to increased absenteeism, labour turnover and, in some cases, industrial unrest. An example of this is a situation where one or several workers cause a line shutdown to create an additional break for relaxation. It is for this reason that much work has been done to enhance line processing so that human factors can be taken into account.

One of the suggested alternatives is to have a number of identical assembly lines and to increase the cycle time and therefore the work content at each station. Some workstations may already require duplication in order to provide the necessary throughput from a slow process, for instance in electroplating, where speeding up the process will lower quality levels drastically. While this maintains throughput, the cost of duplicating all of the equipment for each extra line or subline often proves too expensive. While this would not generally be a viable option in car manufacturing,[2] it is perfectly valid where the labour costs outweigh equipment costs. Another option would be to allow buffer stocks to arise in between adjacent stations, so that additional time is available at the end of each cycle for any necessary adjustments.

The monotony of each workstation can be reduced by producing similar but different products on the same line. Thus the range of tasks for each operator may vary from one part to the next, while maintaining the benefits of mass production techniques.

Cellular manufacture

Cellular manufacturing involves dividing the resources of an organization into discrete units, each of which produces a range of products. This can create a greater sense of teamwork within the cells, which can also lead to increased responsibilities. Where necessary, it allows the traceability of faults, since all the operations on a product or an important subassembly are performed by a small group of workers. There are two different reasons for taking such action, which shall be referred to as "focus" and "group technology".

---

[2] A number of manufacturers have succeeded in implementing such systems by combining line processes with cellular manufacture (e.g. Volvo).

Focus

The idea of focus applies to organizations which produce a range of different products. Production operators in a cell are allowed to concentrate on the selling points of the product rather than general production rules within the factory. Their work can be aimed specifically at one product rather than a number of different products with different demands on production. This method may not help to reduce production costs, but will help to produce a product that is better suited to the customer requirements laid down by marketing. This process of segmentation can also be extended to include many of the support departments, in effect creating separate businesses within the same plant, each with their own staff.

CASE-STUDY 4A

"Focusing" production in an electronic company

An electronic instrument company that made fuel gauges and automatic pilot instruments in the same plant suffered losses on the fuel gauges for several years before the plant manager segregated one part of production from another and even built a physical wall in between them. Production and inventory control were handled separately, as was quality control, and within four months the fuel gauge business had showed profits. This sudden change was due mostly to the relaxation in the control systems for the fuel gauges, which had far looser specifications than the automatic pilot instruments. Care must be taken, however, to avoid focus becoming so important that equally essential features of a good production system are lost, such as synergy and economy of scale.

Source: Skinner, 1985.

Group technology

Group technology is an approach that can be used whenever there is a number of fairly similar products, which can be grouped together and manufactured in one cell. It supposes that the similarities of the parts reduce the time taken for set-ups, reduce tooling and material requirements, while demanding only a limited range of activities from the operators. Thus a number of small batches of products can be combined to make a large batch of similar products, as the cell contains all of the resources necessary to complete each differing product. The main problem is in dividing the available machines and parts into families, as it is almost impossible to produce mutually exclusive groups of the right size. Compromises usually have to be made, and this results in either the duplication of machines, which leads to low machine utilization, or the inter-cell movement of parts during processing, which increases work in progress and transport costs. Although this form of manufacturing can be made to work in most production environments, it is ideally suited to companies where:

- distinct and fairly obvious product families exist

- machines are fairly small and therefore mobile

- many machines have been duplicated.

Possibly the best application of group technology is within light assembly, where the equipment involved is generally hand-held and cheap. Such organizations have the advantage that they can react quickly to changes in product configuration and specification and product mixes by rearranging the cells quickly and cheaply. Where equipment is less mobile or subject to positioning restrictions because of dust extraction, heat, etc., group technology can still offer a number of advantages, but the pace of change will be necessarily slower. What is important when choosing group technology is to create flexibility within the organization. No one machine or operator should be immovably fixed to a cell or its products: a fall in demand in one cell can then allow the resources of that cell to be employed elsewhere. It should be remembered though that, in any such organization, the quality of the workforce is of ultimate importance, as their inherent flexibility can help to overcome the lack of flexibility of the equipment.

Grouping of equipment and people into cells also gives opportunities for improving quality and performance. Each cell can be seen as a little business, receiving material in one state and delivering it elsewhere in another. This gives opportunities for quality improvement, particularly if personnel in the cell can test what they have done.

CASE-STUDY 4B

Group technology at Northrop Aircraft

Northrop Aircraft make a number of small components for the F/A-18 Hornet fighter aircraft, such as brackets, clips, spacers and fillets. They use a group technology approach for this manufacture, with all of the components made under one roof. Before the introduction of this method of manufacture, parts were made in small batches with time scales of five to 12 weeks, with actual processing times of one day. Each part had to travel through more than 16 workstations, spread out over an area of five square miles.

The new system has an 80 x 200 foot production area, containing raw materials, machinery, tooling, computer links and quality assurance. Twenty-five workstations are linked by conveyor belts under computer control. Parts of a similar design are fabricated in the same cell, and the throughput time has been reduced to less than one week. Work-in-progress has been cut by 92 per cent, with reductions in labour and transport costs, and a positive impact on product quality.

Kanban techniques

With the ever-increasing complexity of products, the control of work-in-progress and raw material inventories can become a problem. The optimal situation is to have just enough parts within the system so that it can work efficiently, without having excess stocks lying idle on the shop-floor. Kanban is a very simple technique which is designed to help cope with materials flows throughout a manufacturing system. The basis of the approach is a Kanban card, which displays information relevant to the parts it accompanies. Such information may include:

- quantity of components making up one "Kanban"

- component identity number

- process flow information (source, destination, etc.)

- processing description.

The cards are often fixed to a box, which contains the parts in question and may also be referred to as the "Kanban". When such a box arrives at a workstation, the operator carries out the prescribed action on all of the parts within the box. Once he or she has completed this task, the parts are placed in another Kanban, ready to go to another processing station (see figure 4.4). The empty Kanban is now a signal to the system that this workstation requires more work. In a simple flow system, it may be returned to the previous station, where it would be refilled with more work and brought back, ready for processing. This idea of signalling can be extended throughout the entire production cycle, so that spaces on a finished goods shelf should trigger the production of finished goods, which in turn triggers the release of raw materials and so the purchasing of more stock. This situation helps to keep stock both on the shop-floor and in stores at a manageable level, thus making the production status more clearly visible and helping to reduce working capital.

Despite the simplicity of the Kanban concept, the introduction of such systems can often be a very complicated affair. Perhaps the best way of overcoming the inevitable problems of its introduction is through simulation of the manufacturing process by computer or using manual methods. Paper-based simulation may in some situations be preferable, since it easily shows up problems in the model chosen.

CASE-STUDY 4C

Experimenting with Kanban techniques at 3M, Minnesota

At 3M in Hutchinson, Minnesota, a "coffee cup simulation" was carried out, whereby plastic cups were used to represent Kanban boxes, with pieces of paper inside them as the Kanban cards. A model of the factory was drawn on a large piece of brown paper, accurately mapping the positions of the machines and assembly benches, with set-up times, etc., written nearby. The shop-floor staff were then invited to view a simulation of the system at work and asked to give their ideas on any possible pitfalls such as down time, late deliveries and rework. This led to an extensive debugging period before the system was introduced on the shop-floor. The effects of the simulation were effectively threefold:

- A working system was developed and tested before introduction to the shop-floor, thus causing a minimum of disruption to production.

- When the system was finally introduced, all members of the workforce were fully trained in its use.

- The involvement of the workforce from an early stage meant that the workers felt a sense of ownership for the scheme, and wanted it to succeed as much as the management.

The important lesson here is that the benefits of the "coffee cup simulator" are far greater than those which might be produced using a computer simulation package. While such a high technology method might have produced more accurate results, they themselves are based upon random events, and may not have taken into account all of the points raised by the workforce.

Figure 4.4 The Kanban concept

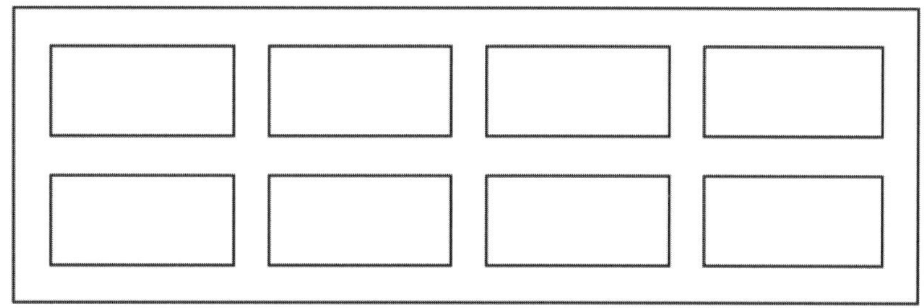

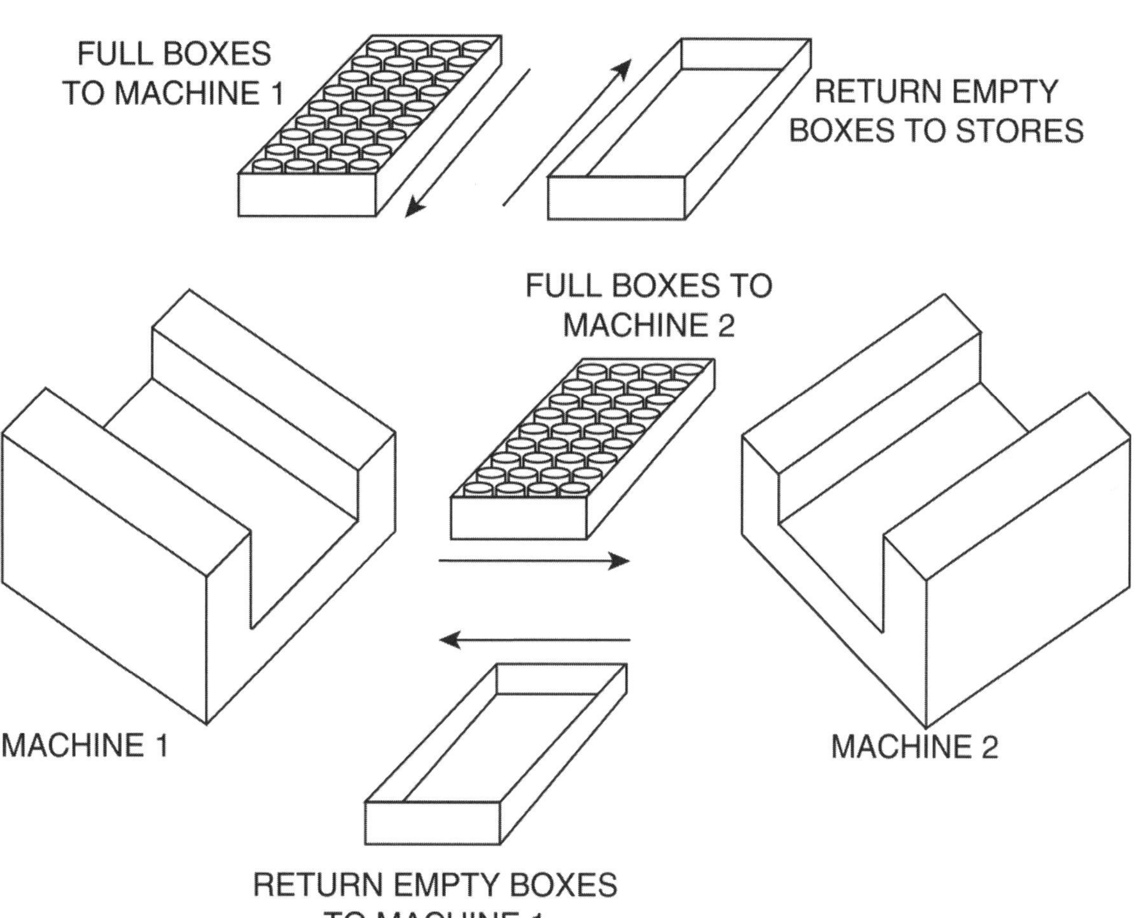

Equipment

The technology available to an organization will, to some extent, determine the form which their operations take. Technological developments, such as surface mounting in electronics, can set a new standard for products, which is soon expected by the customer. However, technological developments have far more wide-ranging consequences than just product development. For example, the use of robotics in car assembly has drastically reduced the number of people required for tasks such as welding and spraying. While this has resulted in the elimination of some physically demanding and unpleasant tasks, it has also had an adverse impact on employment.

A common fallacy is that automation will ultimately reduce the number of people required in production, by replacing people with machines. This is not necessarily the case, as complex machines require skilled programming and expert maintenance and may break down, thus calling for human intervention. Generally, even very flexible automation equipment does not reach the level of adaptability to new tasks of which human operators are capable. This realization has led to a greater emphasis in recent years being placed on the man-machine system and on designing systems with human operators in mind rather than total automation. The productivity and economic viability of advanced technologies are ensured only when tasks have been divided rationally between people and machines, with due regard being paid to human physiology and psychology.

Increased levels of automation in different companies have been linked with the following changes, some of which must be perceived as positive (+) while other effects clearly have a negative (–) impact on workers:

- Increased isolation of workers and reduced interaction (–)

- Reduction in the level of physically strenuous and dangerous tasks (+)

- Pacing of work controlled by machines (–)

- Better working conditions and safety (+)

- Different or new risks (–)

- Increased use of shift work to increase productivity of expensive systems (–)

- Reduction in direct tasks and increase in control or "minding" tasks (+/–)

- Fewer production-related jobs (–)

Thus, the introduction of new technology need not always have a detrimental effect on employment, nor does it necessarily result in less interesting or less demanding jobs. There are alternatives to the dehumanization and deskilling of work, and effort should be concentrated on these aspects, unless some of the mistakes of the first Industrial Revolution are to be repeated.

Table 4.1 gives a broad overview of where people's experience of work, as well as their ability to contribute to better production, can be improved. Each technology provides its own opportunities, but many of those listed across the top of this figure are often used in combination. This opens up even more possibilities, which can be introduced to the general benefit of businesses and their personnel if they are considered at the start of discussions on the introduction of new technologies.

Table 4.1 Main areas of optional choice, by technical system

	NC CNC DNC	Industrial robots	CAD	FMS	Information systems for production (CAM)		CIM	PC	Office automation
					MIS	CAPP			
Work flow				x			x		
Recomposition of tasks	x			x			x		x
Redefinition of machine/worker interface	x	x	x	x	x	x	x	x	x
Programming	x	x	x	x	x	x	x	x	x
Methods of operator instruction	x (DNC only)	x	x	x	x	x	x	x	x
Degrees of operator autonomy			x	x	x		x	x	x

CAD = Computer-aided design

CAM = Computer-aided management

CAPP = Computer-aided production planning

CIM = Computer-integrated manufacture

CNC = Computer numerical control

DNC = Direct numerical control

FMS = Flexible manufacturing system

MIS = Management information system

NC = Numerical control

PC = Personal computer

Source: De Paoli et al., 1986.

Numerical control (NC, CNC, DNC)

The use of computers to control production machines has allowed far greater precision to be obtained from manufacturing processes. This concept originated with NC machining, where programmes were stored on a length of paper tape and fed into a reader mechanism on the machine, instructing the axis drive motors how to move. This soon advanced to computer numerical control (CNC), whereby the machine tool was fitted with a computer control unit. This allowed the loaded programmes (either from paper or magnetic tape) to be altered by the operator, so that changing tool offsets caused by tool wear and regrinding could be taken into account and operations stepped through slowly, if necessary. This technique has now advanced to direct numerical control (DNC), where the tape loader has been replaced by a fixed digital link to a remote computer where all of the programmes are stored on a database. Where such systems exist in modern factories, the operator merely has to dial in the appropriate programme number and set the programme running. Thus the motor and discretionary skills of the worker have become unnecessary for an NC machine operator who is often reduced to a role of "minding the machine".

However, while the operators are less involved in the actual running of the machine, there is more need for them to be able to interpret drawings, set-up sheets and programmes. Thus, the operators need only monitor the machine during its cycle, releasing time for other tasks, which may include quality control or set-up for the next tool or part. There is also the opportunity for the operator to become more responsible for his or her machine, learning skills necessary for routine maintenance, calibration, programming, etc. Automation of machining should not result in deskilling, but should change the type of skills required by the operator, whose skills will complement the scope of the machine rather than supplement it, helping to create an integrated man-machine system like that shown in figure 4.5 and exemplified in case-study 4D.

CASE-STUDY 4D

Integrated man-machine system at Detroit Diesel Allison in Indianapolis

Detroit Diesel Allison, a division of General Motors, makes diesel engines, power shift transmissions and other related products. The plant, employing over 7,000 people, was one of the first in North America to introduce just-in-time principles successfully. An internal manufacturing group known as MAN (materials as needed) organized a survey of the workforce, asking questions related to their work. One question, "What do you think is the major reason for machine breakdown?", was answered as "inadequate lubrication" by around 70 per cent of the workforce.

The group took swift action, engaging the maintenance group in the extensive training of each worker in how to lubricate her or his equipment correctly. Barrels of lubricants were placed at strategic sites on the shop-floor. Such was the success of this action that the maintenance department was later requested to train operators in more extensive repair tasks, such as changing and tightening belts, and replacing seals. Workers felt a greater ownership of their individual machines, and thus accepted responsibility for any forced down time. Any leaks were noticed and acted upon immediately, rather than ignoring the problem and blaming maintenance for any breakdowns.

Figure 4.5   The man-machine environment

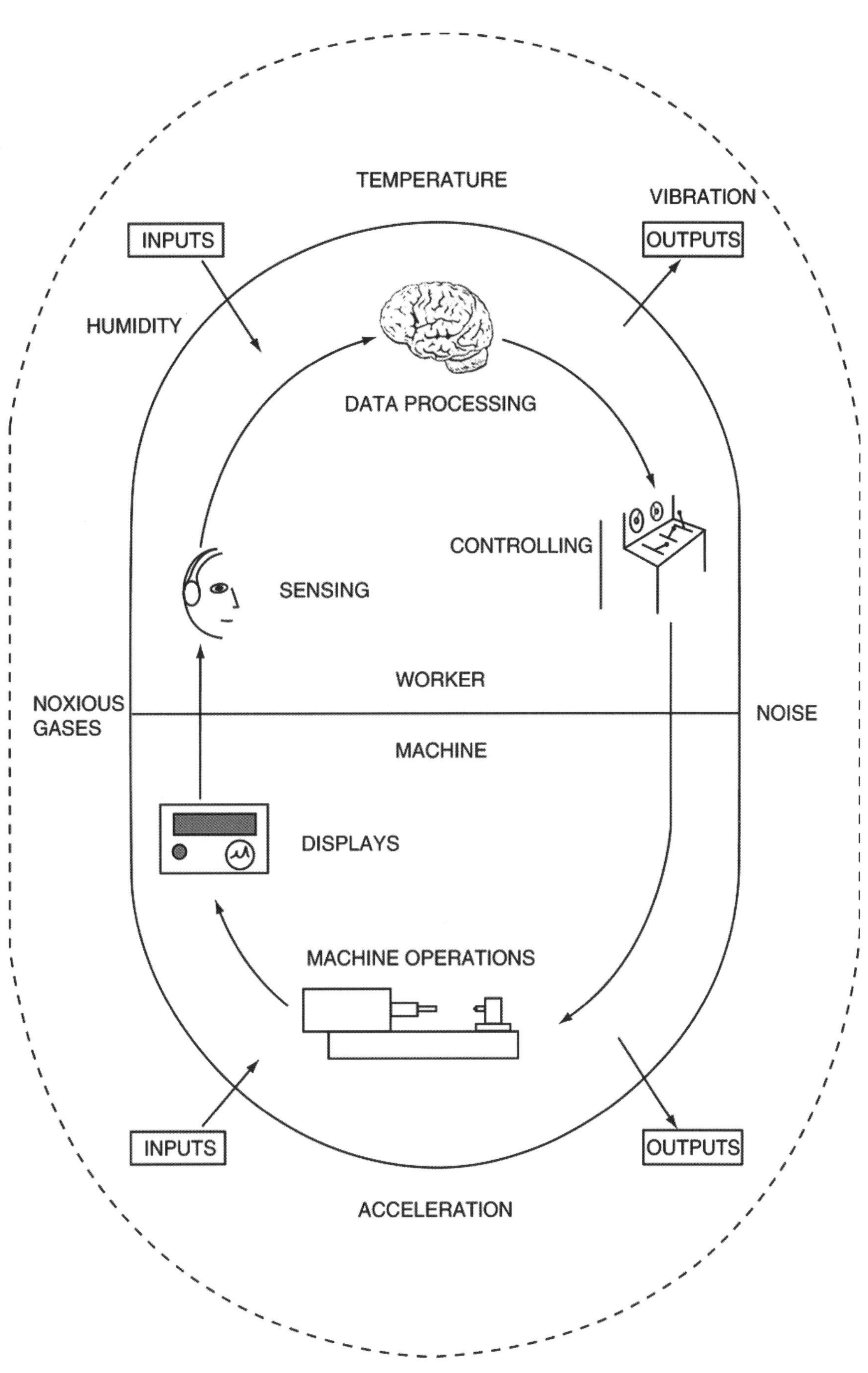

Industrial robots

Industrial robots can be regarded as reliable, consistent and accurate manipulation devices designed to replace people in a production environment. This is, to some extent, a negation of the abilities of people, who are almost infinitely more flexible and intelligent. Robots can be used in high-volume assembly jobs, helping to reduce the cost of the product due to lower assembly costs. They are well-suited to tasks such as machine loading and unloading (e.g. for injection moulding machines, spot welding and seam welding) and finishing operations, such as paint spraying and polishing. All these tasks require good repeat accuracy and a limited degree of flexibility to cope with variety in the product range handled. The magnitude of the fixed costs can negate these advantages, however, as not only are the robots expensive, but the costs of the manufacturing and design engineering involved in setting up an automated robotic assembly line are usually very large. While human assembly workers would increase the unit cost of the product, the fixed costs involved would be very low and, should sales fluctuate, workers can be transferred to other duties. In a human-based production system, the workers can help with solving assembly and quality problems on a group or individual basis to help counter the effects of human error.

The areas in which the benefits of robots are greatest concern tasks that are either hazardous, require extreme physical effort, are carried out in a hostile environment or require high repetitive accuracy. Table 4.2 is intended for use as a guide in determining where robots can have a valid application.

Table 4.2 Robot applications checklist

Checklist criteria for working conditions

Severity of hazard or event			Weighting
A	–	Risk of accidents	5
B	–	Monotony of work	4
	–	Noise	4
C	–	Exposure to dirt and dust	3
	–	Exposure to high or low temperatures	3
	–	Muscular strain	3
	–	Risk of repetitive strain injury	3
D	–	Exposure to oil or grease	2
	–	Exposure to wetness, acids or alkalis	2
	–	Exposure to gases and vapours	2
	–	Exposure to intense or inadequate light	2
	–	Risk of catching cold	2
E	–	Vibrations	1
	–	Need to wear protective clothing	1

Judged level of risk of event occuring:

Scores:	No risk	0	Low risk	1
	Medium risk	2	High risk	3

Multiply weightings by risk scores and add together:

Total score:	< 34	Human system acceptable
	34-67	Redesign of human system necessary
	> 67	Human system unacceptable

It should be noted, however, that it will be easy to find situations where a mechanical application of this method yields a score which is comfortably within the range of human systems but where robots must be employed. An example of such a situation is the handling of nuclear fuel.

A major factor in the use of robots concerns the question of safety. Many of the possible accidents that can be caused by robots are avoided by enclosing the robot and its associated equipment in a cage. This causes problems in that the robot can become a separate entity rather than being integrated into the manufacturing system. This does not mean that the robot should be unprotected, however, but more thought must be applied to allow the flow of work through the system, possibly by the use of conveyors or automatic guided vehicles (AGVs). In situations requiring very rapid or highly accurate operation, a robot may be replaced by dedicated automated machinery, which would be specially built for the task. The flexibility of a robot could not be exploited in this situation due to the low level of product variety, while the dedicated machinery could be designed to work at the very high speeds and levels of accuracy required by the application. A typical example is that of food packaging where low added value goods are handled in high volumes.

Computer-aided design (CAD)

Computer-aided design (CAD) systems have been a part of design offices in industry for a number of years. Their main advantages are often quoted as the speed of design possible, allowing staffing levels to be lowered, and in the analysis capabilities available through associated computer-aided engineering (CAE) software. However, these views are fairly short-sighted, and would probably result in few, if any, productivity or quality improvements. It is true that, after the initial learning cycle, design lead-times will shorten, but CAD should really be seen as a way of enhancing the skill of the designer, rather than producing greater output. The ways in which CAD systems can improve performance in the design office include:

- avoidance of redesigning components (often through strict adherence to formalized "naming" of the product concerned in a database);

- the automatic generation of machine tool programmes for production of individual parts, of mould shapes or robot motions;

- testing of the product using simulation packages;

- efficient logging of changes and modifications to designs;

- more time available for design rather than peripheral activities;

- better initial product designs, performing better and causing fewer problems for manufacturing.

Advanced manufacturing technology (AMT)

Many companies have introduced new technologies as a means of improving productivity, quality and output, only to find that the benefits are far smaller than expected.

Such a reaction has been experienced throughout manufacturing industry, but how can this be avoided? Experts point to five important areas (see box 4.1) where more attention to detail could reap benefits.

As with any large project, planning is of the essence, and can mean the difference between a fruitful project and a disaster. By spending extra time at the design stage to discover possible problems and analyse their solutions, the implementation of a project can be carried out far more smoothly.

The stability of management/workforce relations is also very important when introducing advanced technologies, as such decisions will have considerable effect on the relationship. In many companies where automated machinery was introduced successfully, there were firm believers in the benefits of new systems at all levels of the management hierarchy. While this support may contribute to success, it is not in itself a guarantee of a productive implementation. Another major factor is the attitude of the workforce and their trade unions to the proposed introduction. Managements in many of the more successful companies had involved all those likely to be affected by the implementation at an early stage.

Box 4.1   Important facts in AMT implementation

1. **Include the potential users and managers of AMT in the selection of equipment. They will have a better idea of the particular needs of the company and processes in this area. It may also give them a sense of ownership for the new technology, which may reduce some of the resistance to its introduction.**

2. **Ensure the system will be able to live up to its potential benefits by organizing all of the necessary training for operators and support services. Also ensure that any computer interfaces are easily understandable, and that operator tasks are properly designed.**

3. **Ensure that the new equipment enhances the jobs of the operators rather than de-humanizing them. The new tasks should not be too far removed from their original jobs, but can be enhanced using the techniques outlined in earlier sections. More mentally demanding jobs are often favoured by operators, and may result in greater productivity.**

4. **Ensure that the human aspects of the implementation are effectively managed. This is especially true with regard to industrial relations, and any experience gained can be applied to later technological introductions and other changes to the working environment.**

5. **Remember that learning to be effective with new equipment takes time. If manpower savings are expected, use the initial "surplus" to maintain output while learning is achieved. This applies as much to direct operators as to management, where changes will be needed in what they do.**

Box 4.2 Planning and involvement in AMT implementation

- Examine the precise nature of the business for which the AMT is to be designed. Assess the effect of possible future changes.

- Establish a complete conceptual manufacturing approach for the entire factory/plant/site/business. Establish a suitable facility and systems outline plan.

- Calculate the steps necessary to achieving the goal and the rate at which these are to be attempted and implemented.

- Ensure that the plan is at all times flexible, as it will inevitably change.

- Aim for an integrated system in the long term. Build on a solid database foundation. Go for a well-structured, modular approach.

- Get objective advice but maintain direct responsibility.

- Dedicate as much time to planning the implementation as you would to product development.

- People are the key to success. Involve all concerned at an early stage and keep them informed of progress.

Source: Adapted from Kirton, 1986.

Tooling-based solutions

As the previous section highlighted, leading edge technology (or hardware) is not always the solution to problems in industry. There are many ways in which a company can improve its ability to compete without spending huge sums on the latest technology. Such techniques can be described as Advanced Manufacturing Methods (AMMs).

Reduction of set-up times

For many years now, it has been reasoned that components can only be manufactured in certain volumes (Economic Batch Quantities or EBQ) in order to minimize set-up costs. This inevitably leads to large levels of unwanted stock being produced rather than those items that could actually be sold to the customer. The whole idea of economic batch quantities is put into question when the storage and handling costs of unwanted semi-finished and finished items is taken into account, not to mention the time lost on the machines which could have been used to make saleable products. The important thing to remember is that the cost of manufacturing components is not affected merely by the volume produced, but also by the time taken to set up the job. If set-up times were reduced to zero, then an organization could produce an infinite variety of products as efficiently as a dedicated flow line. Quick changeover is the foundation for successful manufacture of high variety, low lead time products.

Many large machining centres have access to a wide variety of cutting tools stored in some form of rack. The tool holders are specifically designed so that they can be quickly loaded and unloaded by the machine as it changes from one cutting operation to the next, helping to keep machining times to a minimum. Some machining centres even operate using quick-change pallets so that the work piece set-up time is also kept as low as possible. Similar techniques and methods can be applied in workplace design at a more general level.

By reducing set-up times on machines, tooling costs for products can be reduced and product lead times can be cut. This implies faster response to product changes, allowing manufacturing to be more flexible in terms of product mixes.

Change-over time reduction

The first step towards change-over time reduction is to analyse the current change-over procedures, splitting the operations into internal and external actions (see figure 4.6). Internal actions are those that can only be performed while the machine is stopped (e.g. tool changes), while external actions can be done while the machine is still running (e.g. gauge calibration). Once this has been done, the internal actions must be carefully examined to see if there is any scope for converting them into external actions. There are several ways in which this can be achieved, and examples include quick change-over tooling, standardized locating plates, use of common datums and improved component design. Once this stage is complete, effort should be focused on streamlining both external and internal operations. External actions should never be carried out while the machine has stopped, as this represents a waste of potential machining time. As the emphasis lies on

Figure 4.6   The four steps to set-up time reduction

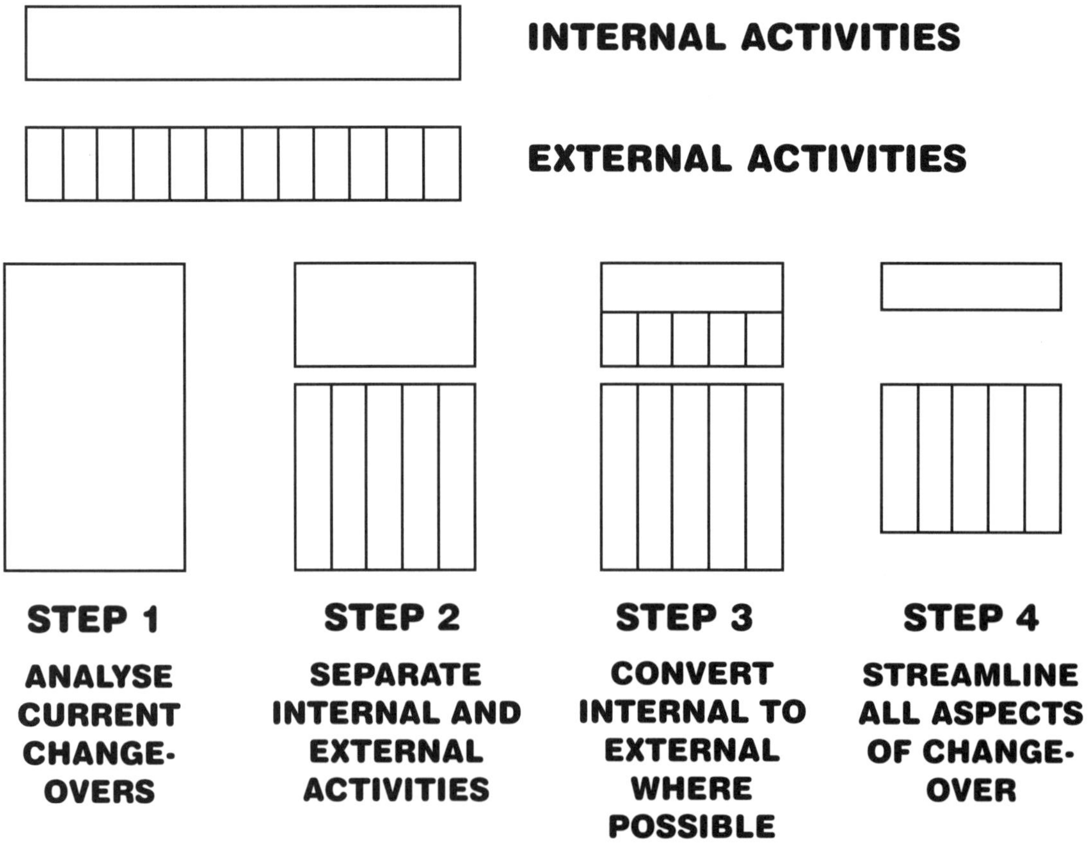

continuous improvement, pressure should be maintained to reduce internal actions as far as possible, while maintaining external actions at a manageable level. Companies that have examined their operations in this way have been found to reduce their change-over times by around 60 per cent.

The effects of faster change-over times can be more far-reaching, however. Once the idea of economic batch quantity is no longer relevant, parts will be produced in far lower volumes. This helps to reduce the level of capital employed in finished or semi-finished inventories and, at the same time, the amount of storage space required and the cost of handling the parts. The reduced lead times on products also mean that work spends less time in the system, and so work-in-progress (WIP) levels should fall, as long as current processing levels are maintained. In the past, companies have tended to keep WIP levels fairly high, so that any deficiencies in the system could be cushioned. However, since the deficiencies are exactly the thing that we are trying to control or eliminate, lowering WIP levels gradually will help to identify areas where improvement is necessary. The analogy most commonly used is that of the rocky river bed, where the rocks can only be found when the water level (or stock level) is reduced (see figure 4.7). Once these rocks have been removed, water levels can be dropped further still.

Figure 4.7   Identification of problem areas by reducing work-in-progress (WIP)

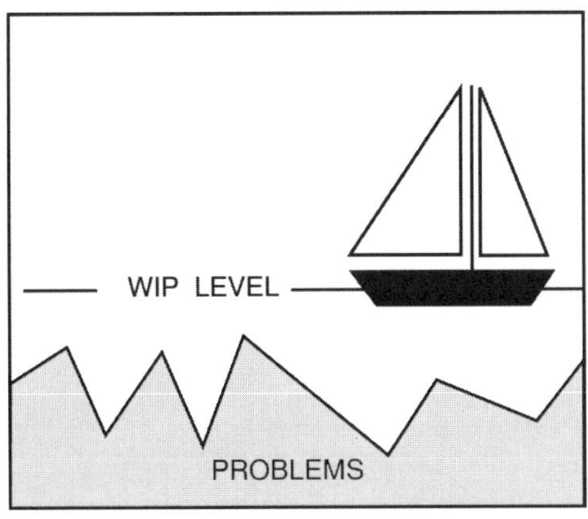

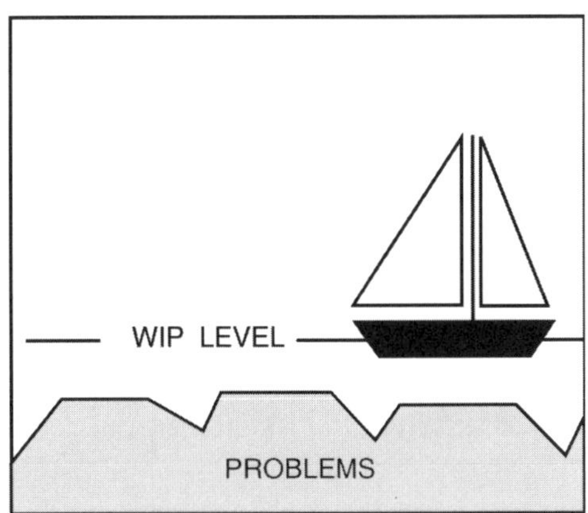

The lesson to be learned here is that solutions to productivity and competitiveness problems need not come from the latest technology. It is better to add small, relatively simple machines one by one as demand increases than to rush into a total system upgrade, which will cause many months of down time before any benefits are gained. Smaller machines are easier to maintain, and more machines can provide flexibility in the product mix and offer protection against machine breakdown. Having a number of identical machines also allows them to be spread throughout the manufacturing plant, possibly having one per production area. There is no need to replace such equipment unless either it ceases to cope with the tasks required (despite efforts from maintenance) or when production technology is replaced by innovation. Such equipment expenditures apply equally to new products as to existing markets. Case-studies 4E and 4F give good examples of how high technology will not of itself provide an advantage and how progressive introduction and discrete use of technology can often lead to better results.

CASE-STUDY 4E

"Discrete" use of technology at Toyota plant

Toyota's No. 9 Kamigo engine plant is equipped with machines that, in some instances, are 20 years old. Over the years, they have been consistently retrofitted and upgraded using limit switches and electric eyes. If a machine breaks down or produces a bad part, then a giant overhead screen informs the appropriate staff, helping to keep levels of quality and output as high as possible. The plant runs only two shifts, with preventive maintenance between shifts and during lunch breaks. Machine set-up times are low, so batch sizes are small and product variety is high. The low WIP means that there are no shop-floor storage areas, and so machines are closely situated to each other, allowing one operator to control several machines. The provision of automatic loaders and setters also means that the operators' duties can be carried out fairly quickly.

The managers at Kamigo consider that having 20-year-old equipment merely means that they have 20 years of experience at getting the machines to work efficiently.

CASE-STUDY 4F

"Discrete" use of technology at the Tektronix plant in Oregon

In 1982, the Tektronix plant in Oregon started production of a new precision video display product. Market surveys had predicted that the product would be a big hit in the market-place, and consequently Tektronix had geared production for high volumes. As quality and lead times were major factors in how well a product sold, no expense was spared in the creation of the production line, including the purchase of top quality components and the building of three large clean rooms. The production flow was designed to reduce delays and keep shop-floor inventories to a minimum. The high costs incurred by this initial outlay was expected to be offset by the low unit costs of the product.

The forecast sales predicted that the facility would be operating three shifts by the middle of 1983. The reality was that the line was running for less than half a shift, as the product failed in the market-place. Some of the resources could be moved to other areas, but the majority, including the clean rooms, lay idle. This lesson was not learned in vain, however, and when another product, a colour-shutter, was ready for market introduction, more care was taken in production.

The plan this time was to increase gradually the levels of production. Initial equipment purchases would be followed by further orders, but if the product sold in low volumes then the deliveries would be slowed down or cancelled altogether. Assuming that the product sold well, then the production capacity was to increase gradually each time a new set of equipment arrived and was put into action.

This approach had a number of advantages over the establishment of one large production line. First, the experience gained from the early flow lines could be used in the organization of each of the following lines, improving the process where necessary. Equipment failures in one line were of less importance, as work continued on other lines, and the affected staff could be diverted elsewhere within the system as their skills were relevant to all lines. Where a product had a number of different models, these could be produced simultaneously, either on different lines, or on the same line if equipment change-over times were short enough. This, in turn, meant that marketing did not have to predict exact ratios of product mix, as some of each product was being made each day. The product mix could be changed at fairly short notice, as production never stopped on all of the lines simultaneously, allowing finished goods inventories to be kept to a minimum.

Layout

Production layout methods vary to some extent, depending on the type of production flow used. The underlying methodology remains the same, however, and some basic objectives for an efficient layout can be clearly defined.

The main criterion in layout design is to keep the transfer of raw materials and semi-finished goods between workstations to a minimum. This will reduce transportation costs, throughput time and work-in-progress (WIP), and can thus reduce costs and increase productivity. Areas that have frequent interaction or require movement of large and cumbersome objects should be positioned in close proximity to each other. As there are likely to be many such interactions, the final layout is produced by a process of compromise.

While it is desirable to reduce buffer stocks and transfer of goods between workstations, it is also important to reduce the levels of WIP which, in turn, reduces the need for shop-floor storage areas. This tends to result in very compact factory layouts, allowing machines to be moved closer together. This also gives a much clearer view of how the processes are working, allowing any bottlenecks to be identified and managed.

One of the most important things to realize when designing a layout is that any design will have a finite lifetime. In some cases, it will cease to be efficient after a matter of months. As product mixes, volumes and designs change, and with the introduction of new machines or technologies, the loading placed on each workstation will also change. This is a very important factor to bear in mind when choosing a layout, and implies that the optimal layout is not always the best. The important aspects of future system performance that should be taken into consideration are:

- current and predicted demand for the product(s);

- the current capacity and capacity required for the predicted demand;

- how any shortfalls in capacity can be handled;

- the space requirements for future expansion;

- how future products will fit into the system.

The following sections deal with some of the analytical methods that can be used for layout planning. Many software packages are available on the market for such planning.

The sequencing of production facilities

Cross charts

Sometimes called "from-to" charts, this method of facilities planning tallies the number of movements between areas or departments in the processing of a product. Where materials handling costs are high between departments, possibly arising from the need to use a fork-lift truck, this can be taken into account in the appropriate entry on the chart. If a current facility exists, then the entries could represent the product of materials handling cost, distance and frequency of trips. As the sum of the individual costs represents the total materials handling cost, this can be used to evaluate alternative designs, where the different values of distance can be substituted. In the case of new facilities, the departments with more frequent trips should be positioned near to each other.

In the example of table 4.3, the numbers could represent the quantity of products or number of pallet loads to be transferred from the departments listed on the left to those listed along the top. The departments (areas or machines) are set out such that distances (or other transport "costs") between adjacent departments are least. Thus it is more costly to transport from department 2 to department 5 than to department 3. Hence, the further from the diagonal, the more the cost.

Table 4.3(a)   An example of "from-to" chart: Situation before change

To department

From department	1	2	3	4	5	6	7	8
1		75	100	90	40		30	60
2			100		150			
3					70		30	95
4					90	70		100
5			70			20	50	30
6					30		100	30
7						30		120
8								

Cost function: $(75+100+100+70+90+70+20+50+100+30+120) \times 1 \; + \; (90+40+150+30+100+30) \times 2 \; + \; (30+95) \times 3 \; + \; 60 \times 4 \; + \; (70+30+30+) \times 1 \times 1.5 = 825+880+375+240+1950 = 2515$

Table 4.3(b)   An example of "from-to" chart: Situation after change

To department

From department	1	2	3	5	4	6	7	8
1		75	100	40	90		30	60
2			100	150				
3				70			30	95
4			70			70		100
5				90		20	50	30
6					30		100	30
7						30		120
8								

Cost function: $(75+100+100+150+70+70+20+50+100+30+120) \times 1 \; + \; (40+90+30+100+100+30) \times 2 \; + \; (30+95) \times 3 \; + \; 60 \times 4 \; + \; (70+90+30+30+) \times 1 \times 1.5 = 885+580+375+240+330 = 2410$

Any numbers above the diagonal move along the sequence, those below it will be backtracking. By manipulating the position of departments in the sequence, the numbers running just above and along the diagonal are maximized. Thus minimum movement cost and maximized forward flow are achieved. The sum of the number in each cell multiplied by the distance from the diagonal (with a further factor introduced for any numbers below the diagonal to represent the costs of backtracking) is the cost of material movement for the system. As shown in table 4.3(a) and (b), it is possible to reduce the cost of internal transport by exchanging the locations of departments 4 and 5. As a matter of course, such a move can only take place if there are no overriding considerations which necessitate a particular arrangement of the factory. Further improvements may well be possible but are left to the reader as an exercise.

Balancing the production sequence

The agreed cycle time provides a basis for designing the flow line. Figure 4.8 is a precedence diagram for the assembly of a product, with each numbered circle representing a separate assembly stage, and the numbers outside the circle corresponding to the duration of the action. Consecutive production stages are represented by columns numbered I, II, etc. Each stage can only be completed once all of the stages feeding it (represented by incoming arrows) have been completed. A definite order can also be established, whereby those elements (or tasks) in column I can be started immediately, while those in column II can only be started after one or more activity in column I has been completed, and so on. This information is then best represented in tabular form (see table 4.4), which shows where certain stages have possible membership of more than one precedence column. It should be pointed out that stages of production do not necessarily mean that for each an independent workstation is needed.

A fuller description of this example is given below.

The Kilbridge and Webster method

The purpose of this method is precisely to ascertain the number of workstations needed for assembling a product and what elements each will assemble. Let us suppose that assembly of a simple component requires the performance of 21 work elements which are governed by certain precedence constraints, as shown in figure 4.8. This

Figure 4.8 Precedence diagram of work elements

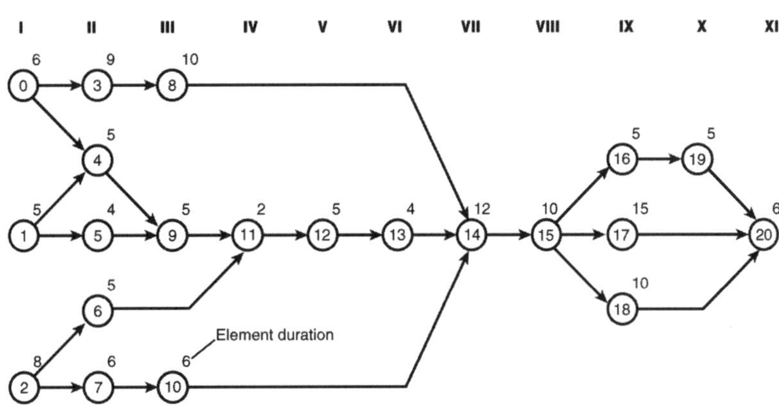

Source: Adapted from Wild, 1989.

precedence diagram shows circles representing work elements placed as far to the left as possible, with all the arrows joining circles pointing to the right. The figures above the diagram are column numbers. Elements appearing in column I can be started immediately, those in column II can be begun only after one or more in column I has been completed, and so on.

The data shown on this diagram can now be represented in tabular form (see table 4.4). Column (a) represents the production stages; column (b) shows the different elements in each stage. Column (c) describes the lateral transferability of elements among columns; for example, element 6 can be performed in column III as well as in column II without violating precedence constraints. Element 8 can also be performed in any of the columns IV to VI, as can element 10. Element 3 can also be performed in any of the columns III to V, provided element 8 is also transferred, as can element 7. Column (d) shows the duration of each element. Columns (e) and (f) are self-explanatory.

Suppose it is our objective to balance the assembly line for an agreed cycle time of 36 units of time. In this case we would proceed as follows:

1. Is there a duration in column (f) of the table equal to the cycle time of 36? *No.*

2. Select the largest duration in column (f) less than 36, i.e. *19 for column I.*

3. Subtract 19 from 36 = 17.

4. Do any of the elements in column (II), either individually or collectively, have a duration of 17? *No, the nearest is 16 for elements 4, 6 and 7, as follows: 5 units + 5 units + 6 units = 16 units. This will give a total workstation time of 19 units + 16 units = 35 units for station 1.*

5. Select the smallest duration from column (f) which is larger than 36, i.e. 48 *for columns plus I + II.*

6. Can one or more of the elements in columns I and II be transferred beyond II so as to reduce the duration as near as possible to 36? *No, but element 3 (with 8) plus 6 can be transferred to give a workstation time of 34.*

7. Select the next largest duration from column (f), i.e. *69 for columns I + II + III.*

8. Can one or more of the elements in columns I, II and III be transferred beyond column III so as to reduce the duration of 36? *No, the nearest elements, 3, 7, 8 and 10, would give a duration of 38, which is too large.*

9. Will an improved allocation of elements for this station be obtained by considering a large duration from column (f)? *No.*

10. Adopt the best allocation found previously, *i.e. step 4, which gave a workstation time of 35.*

11. Rewrite the table to show this allocation and calculate new cumulative figures for column (f) (table 4.5a).

12. Is there a duration in column (f) of the new table to equal 36? *Yes, for columns III and IV.*

13. Allocate the elements in these columns to the second workstation and redraw the table showing new figures for column (f) (table 4.5b).

14. Is there a duration in column (f) of the new table equal to the cycle time of 36? *No.*

15. Select the largest duration in column (f) which is less than 36. *It is 31 for columns V, VI, VII and VIII.*

16. Subtract 31 from 36 = 5.

17. Does one or more of the elements in the next column (IX) equal 5? *Yes, element 16.*

18. Allocate the columns concerned and element 16 to the third workstation and redraw the table (table 4.5c).

19. Is there a duration in column (f) of the new table equal to 36? *Yes, for columns IX, X and XI.*

20. Allocate the element in these columns to the fourth workstation.

All 21 elements have now been assigned to form workstations in the manner shown in figure 4.9.

Table 4.4   Tabular representation of data in figure 4.8

Column no. in precedence diagram (a)	Element no. (b)	Transferability of element (c)	Element duration (d)	Duration of column (e)	Cumulative duration (f)
I	0		6		
	1		5		
	2		8	19	19
II	3	III-V (with 8)	9		
	4		5		
	5		4		
	6	III	5		
	7	III-V (with 10)	6	29	48
III	8	IV-VI	10		
	9		5		
	10	IV-VI	6	21	69
IV	11		2	2	71
V	12		5	5	76
VI	13		4	4	80
VII	14		12	12	92
VIII	15		10	10	102
IX	16		5		
	17	X	15		
	18	X	10	30	132
X	19		5	5	137
XI	20		6	6	143

Table 4.5(a)   First stage of conversion

Column no. in precedence diagram (a)	Element no. (b)	Transferability of element (c)	Element duration (d)	Duration of column (e)	Cumulative duration (f)
I	0		6		
	1		5		
	2		8		
II	4		5		
	6		5		
	7		6		(35)
III	3	III-IV (with 8)	9		
	5		4		
	9		5		
	10	IV-VI	6	24	24
IV	8	V-VI	10		
	11		2	12	36
V	12		5	5	41
VI	13		4	4	45
VII	14		12	12	57
VIII	15		10	10	67
IX	16		5		
	17	X	15		
	18	X	10	30	97
X	19		5	5	102
XI	20		6	6	108

(Cumulative duration column, rows I–II: "Station 1")

Table 4.5(b)   Second stage of conversion

Column no. in precedence diagram (a)	Element no. (b)	Transferability of element (c)	Element duration (d)	Duration of column (e)	Cumulative duration (f)
	0				
	1				
	2				
	4				
	6				
	7			35	(35) Station 1
III	3		9		
	5		4		
	9		5		
	8		10		
	10		6		
IV	11		2	36	(36) Station 2
V	12		5	5	5
VI	13		4	4	9
VII	14		12	12	21
VIII	15		10	10	31
IX	16		5		
	17	X	15		
	18	X	10	30	61
X	19		5	5	66
XI	20		6	6	72

Table 4.5(c)   Final stage of conversion

Column no. in precedence diagram (a)	Element no. (b)	Transferability of element (c)	Element duration (d)	Duration of column (e)	Cumulative duration (f)	
	0					S
	1					t
	2					a
	4					t i o n
	6					
	7			35	(35)	1
	3					S
	5					t
	9					a t
	8					i o n
	10					
	11			36	(36)	2
V	12		5			S
VI	13		4			t a
VII	14		12			t i
VIII	15		10			o n
IX	16		5			
				36	(36)	3
IX	17		15			
	18		10	25	25	
X	19		5	5	30	
XI	20		6	6	36	

Figure 4.9   Allocation of 21 elements to 4 workstations

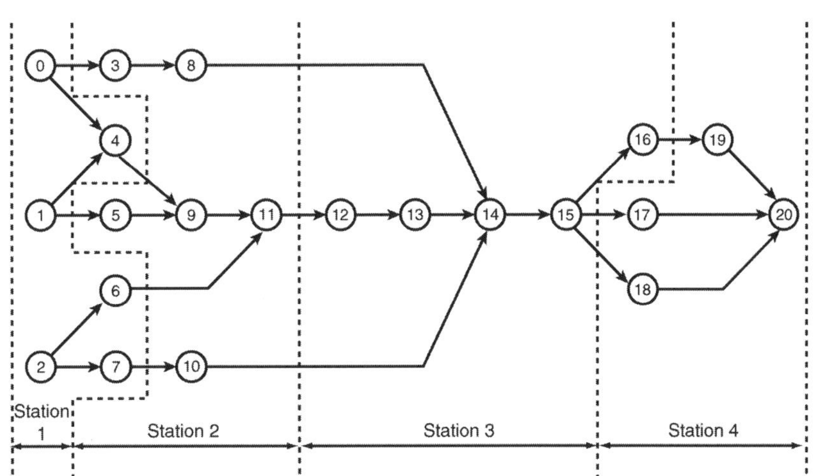

Source: Adapted from Wild, 1989.

As can readily be seen from the example, this method is rapid, easy and often quite efficient. The allocation of elements is basically determined by precedence relationships, lateral transferability of elements being used to aid allocation when necessary. The originators of this method offer the following comments to aid in the application:

(a)   Permutability within columns is used to facilitate the selection of elements (tasks) of the length desired for optimum packing of the workstations. Lateral transfer ability helps to deploy the work elements (tasks) along the stations of the assembly line so they can be used where they best serve the packing solution.

(b)   Generally the solutions are not unique. Elements (tasks) assigned to a station which belong, after the assignment is made, in one column of the precedence diagram can generally be permuted within the column. This allows the line supervisor some leeway to alter the sequence of work elements (tasks) without disturbing optimum balance.

(c)   Long-time elements (tasks) are best disposed of first, if possible. Thus, if there is a choice between the assignment of an element of duration, say 20 units, and the assignment of two elements of duration, say 10 units each, assign the larger element first. Small elements are saved for ease of manipulation at the end of the line. The situation is analogous to that of a paymaster dispensing the week's earnings in cash. He will count out the largest bills first. Thus, if the amount to be paid to a worker is $77, the paymaster will give three $20 bills first, then one $10 bill, one $5 bill and two $1 bills in that order.

(d)   When moving elements laterally, the move is best made only as far to the right as is necessary to allow a sufficient choice of elements for the workstation being considered.

In reality, exact line balancing is rarely possible, often due to restricted cycle times, machine output and zoning effects. Zoning, requiring the proximity or remoteness of two or more adjacent tasks, can be either positive or negative. Precise balancing can be undesirable, especially in manual assembly where the variability of workers' pace can negate the careful work of the system designer. These variables can be smoothed out by making cycle times longer, thus allowing the peaks and troughs of activity to cancelled out. This lowers output and increases the zoning effects, but it can increase the levels of output from operators.

Another very important aspect in flow-line design, especially relevant in the case of assembly lines, is to consider the working conditions under which the operators will be placed. Production lines are often associated with short cycle times and, where this is the case, operators will generally suffer from the tedium of repeating simple tasks throughout the day. These effects can be reduced somewhat by correct design of the line, allowing a degree of interaction between operators, for example, with the use of assembly teams. Many companies now approach production line balancing with some caution, as it assumes that assembly operations take a fixed length of time to perform. As this is not the case in reality, many of the calculated benefits of production lines are diminished or disappear altogether. It is for this reason that many companies now employ some of the human-centred manufacturing techniques of job improvement described in Chapters 3 and 5.

Group technology clustering

The basis of group technology (GT) production lies in the division of products and resources into mutually exclusive cells, with each cell making a limited range of products. This situation is rarely achievable in practice, and some parts will often require processing in more than one cell. The problem can be further complicated where there is a constantly changing range of product lines and mixes, turning perfect groupings into an unmanageable conglomerate. Many methods exist to aid the formation of GT cells, which vary greatly in their complexity, the most popular of these being the matrix-based and linear programming methods.

Matrix clustering

This approach to clustering relies on a two-dimensional incidence array to represent the problem. This array has each column named after a product and each row named after a machine, with a '1' entry in the matrix where a part is processed on a given machine, with a '0' or blank entry otherwise as in table 4.6(a). This matrix is then manipulated, (mathematically), until distinct clusters become visible as shown in table 4.6(b).

Table 4.6   Matrix clustering

(a) PARTS

MACHINES:

	P1	P2	P3	P4	P5	P6	P7	P8
M1		1	1		1			
M2	1					1		
M3				1			1	
M4	1					1		
M5			1		1			1
M6				1				
M7		1	1		1			1

(b) PARTS

MACHINES:

	P3	P5	P8	P2	P1	P6	P4	P7
M7	1	1	1	1				
M5	1	1	1					
M1	1	1		1				
M2					1	1		
M4					1	1		
M3							1	1
M6							1	

Linear programming methods

The benefits of using linear programming methods to solve allocation problems have long been recognized, and several clustering models have been created for the purpose of cell formation. These methods try to measure the current system performance and then improve on this result by carrying out certain defined manipulations. Other mathematical procedures exist for the solution of these problems. Clustering can be done using fuzzy mathematics while expert systems also exist for their solution. These are fully discussed in texts on facilities design, production planning or industrial engineering, some of which are given in the references.

Human factors

While the previous sections described heuristic methods for the design of production systems, little account is taken in these methods of the human factors involved in workplace design. Unless these factors are taken into consideration, both quality and productivity will suffer.

Performance will not be ensured by merely balancing an assembly line or buying the latest information technology, but by ensuring that the people who will be engaged in working with these systems have conditions which make it worthwhile for them to do their work well. While this chapter has concentrated on the technical aspects of planning production flow, it is not to be read in isolation, but together with other chapters, thus emphasizing the importance of a "human working system", one which can match facilities to people so that they are both effective together.

References

J.R.S. Bailey: *Job design and work organization* (Englewood Cliffs, New Jersey, Prentice Hall, 1983).

—: *Managing people and technological change* (London, Pitman, 1993).

D.D. Bedworth and J.E. Bailey (eds.): *Integrated production control systems, management, analysis, design*, 2nd edition (New York, John Wiley, 1987).

J.R. Bessant: "Flexible organisations for flexible manufacturing", in *Automation*, March-April 1989.

V. Bignell et al. (eds.): *Manufacturing systems, context, applications and techniques* (Oxford, Blackwell, 1985).

E.S. Buffa and R.K. Sarin (eds.): *Modern production management* (New York, John Wiley, 1987).

A. Cherns: "Problems, prospects and the state of the art", in *The Quality of Working Life*, Vol. 1 (New York, 1975).

C. Chu and J. Hayya: "A fuzzy clustering approach to manufacturing cell formation", in *International Journal of Production Research*, Vol. 29, No. 7, 1991, p. 1475.

C.W. Clegg and J.M. Corbett: "Psychological and organizational aspects of computer-aided manufacturing", in *Current Psychological Research and Reviews*, Vol. 5, 1986. pp. 189-204.

P. De Paoli, A. Fantoni and G. Miami: *Participation in technological change: Consolidated report* (Dublin, European Foundation for the Improvement of Living and Working Conditions, 1986).

F. Emery and E. Thorsrud (eds.): *Democracy at work: The report of the Norwegian Industrial Democracy Program* (Leiden, Netherlands, Martinus Nijhoff, 1976).

S. Fritz, F. Schmid and B. Wu (eds.): *Manufacturing systems design: A survey*, Computer Integrated Manufacturing Conference, Keele University, United Kingdom, September 1993.

K. Gill: "Human shaping of social innovation", in *Proceedings of the NATO Advanced Study Institute "People and computers — Applying an anthropocentric approach to integrated production systems and organisations"* (Heidelberg, Springer, 1994).

E. Goldratt and J. Cox: *The goal* (Aldershot, United Kingdom, Gower, 1990).

M. Hamlin: *Human-centred CIM systems*, ESPRIT Project 1217(1199) Deliverable 27: Final Report (Hemel Hempstead, BICC Technologies, 1989).

E.W. Housh: "Office furniture's effect on employees", in *The Office*, January 1988, p. 113.

R.G. Jacobs: "OPT uncovered", in *Industrial Engineering*, October 1984.

G. Kanawaty: *Introduction to work study* (Geneva, ILO, fourth revised edition, 1992).

J.R. King: "Machine-component group formation in production flow analysis", in *International Journal of Production Research*, Vol. 18, No. 2, 1980, p. 213.

J.A. Kirton: "Implementing advanced manufacturing technology: The way it is", in C.A. Voss (ed.): *Managing advanced manufacturing technology* (Bedford, United Kingdom, IFS Publications, 1986).

B. Kleiner and P.A. Rudolph: "Building an effective office space", in *Work Study*, Vol. 39, No. 3, May-June 1990, pp. 6-10.

A. Kusiak: "EXGT-S: A knowledge based system for GT", in *International Journal of Production Research*, Vol. 26, 1988, p. 887.

—: *Intelligent manufacturing systems* (Englewood Cliffs, New Jersey, Prentice Hall, 1990).

— and W. Chow: "An algorithm for cluster identification", in *IEE Transactions on Systems, Man and Cybernetics*, Vol. SMC-17, No. 4, 1987, p. 696.

Lucas Engineering and Systems: *The Lucas manufacturing systems engineering handbook* (London, Institution Production Engineers, 1988).

McKinsey Global Institute: *A comparative study of industrial productivity in six key nations* (New York, 1993).

R. Nagi, J. Proth and G. Harhalakis: "An efficient heuristic in manufacturing cell formation", in *International Journal of Production Research*, Vol. 28, No. 1, 1990, p. 185.

—, — and —: "Multiple routings and capacity considerations", in *International Journal of Production Research*, Vol. 28, No. 12, 1990, p. 2243.

W.J. Paul and K.B. Robertson: *Job enrichment and employee motivation* (Aldershot, Gower, 1970).

Rationalisierungs-Kuratorium der Deutschen Wirtschaft: *Productivity and the future of work*, proceedings (Munich, October 1986).

U. Rembold, B.O. Nnaji and A. Storr (eds.): *Computer-integrated manufacturing and engineering* (Wokingham, United Kingdom, Addison Wesley, 1993).

F. Schmid: "The industrial co-operative: An organisational model for the management of advanced manufacturing technology", in C.A. Voss (ed.): *Managing advanced manufacturing technology*, proceedings of the UK Operations Management Association Conference (Bedford, United Kingdom, IFS Publications, 1986).

—: "The human-centred approach to the implementation of computer-integrated systems", in F. Schmid and T. Hancke (eds.) *Control and dynamic systems, Vol. 61: Computer-aided manufacturing — Computer-integrated manufacturing (CAM/CIM)* (San Diego, Academic Press, 1993).

R.G. Schroeder: *Operations management: Decision-making in the operations function*, 4th edition (New York, McGraw-Hill, 1993).

R.S. Schuler, L.P. Ritzman and Davis: "Merging prescriptive and behavioural approaches for office layout", in *Journal of Operations Management*, Vol. 1, No. 3, 1981, pp. 131-142.

W. Skinner: *Manufacturing: The formidable competitive weapon* (New York, John Wiley, 1985).

N.C.D. Slack: *The manufacturing advantage: Achieving competitive manufacturing operations* (London, Mercury, 1991).

C.A. Voss (ed.): *Managing advanced manufacturing technology*, proceedings of the UK Operations Management Association Conference (Bedford, United Kingdom, IFS Publications, 1986).

T.D. Wall, C.W. Clegg and N.J. Kemp (eds.): *The human side of advanced manufacturing technology* (Chichester, United Kingdom, John Wiley, 1987).

R. Wild: *International handbook of production and operations management*, 4th edition (London, Cassell Educational Limited, 1989).

M. Woodward: "Get set ... Go!", in *Manufacturing Engineer*, Vol. 70, No. 6, July-August 1991, pp. 39-41.

B. Wu: *Manufacturing systems design and analysis: Context and techniques* (London, Chapman & Hall, 1992).

WORKING TOGETHER: FLEXIBLE WORK GROUPS AND MULTI-SKILLING

5

David Buchanan

The organizational challenge

So far we have described the work people do, looking at it from the ergonomic perspective of how we design jobs and workplaces to give people the best opportunities to do their work. We then went on to consider how to lay out these individual workplaces to achieve a factory or office, i.e. somewhere where products were produced or from which services were provided.

In the previous chapter, apart from brief comments at the end, it was implied that these arrangements were primarily technical, a matter of lines, conveyors or groupings of relevant machines. But the case-studies indicated that there was more to it than this, and the last section, on intangible factors, was quite explicit on this point.

So, in this chapter we will describe how these arrangements of workplaces can achieve efficient performance, which is to say performance which is of good quality, on time, adaptive to the customers' needs and undertaken by people who are interested in doing it well.

Of course, we would all like this, but you may wonder if you need it, you may believe that things are all right as they are. But does your enterprise have to deal with frequent changes in product designs and specifications, in customer requirements, in the technology of the business, or in materials? Are you coming under more pressure to deliver on time, to the customer's specific requirements? Do you find that you have to be more flexible in your response to customer requests? If so, then you are competing with other suppliers of your goods and services on four fronts simultaneously: cost, quality, flexibility and time.

If these competitive factors apply to your organization, then you should be considering the development of multi-skilled teamwork to help achieve improvements in productivity, customer satisfaction, flexibility and responsiveness to your market. Team-based approaches to the design of work systems also contribute significantly to job satisfaction, skills growth, and to the quality of working life. Competitive strategy is no longer a simple choice between low cost or high quality. Increasingly the competitive organization is one which can offer low cost *and* high quality while being able to respond in a flexible way to market pressures and deliver on time, ahead of competitors. Improvements in work system design thus also lead to improved job security and promotion opportunities for employees.

But why groups? And why multi-skills and teamwork? In the previous chapter, we described how to design technical arrangements to reduce work in progress, to increase the opportunity to change the products quickly, increase the flexibility of equipment use by grouping processes into cells relative to the manufacture of similar products or for the production of complete items, and so on. What we must now do is describe how to arrange for people to be able to use these technical arrangements well, and how to create situations where using them is interesting, involving, and enables people to enjoy the authority and responsibility of being part of a worthwhile whole. That is why we need multi-skills and working groups.

Away from the traditional approach

Traditional approaches to the design of work systems – based on work fragmentation – are not appropriate for organizations operating in a rapidly changing competitive environment. Fragmented tasks are simple and repetitive, and boredom leads to carelessness and lack of attention. Work settings designed in this way are typically prone also to abnormally high levels of poor timekeeping, absenteeism, and to high labour turnover. The loyalty and commitment of the workforce are thus not likely to be high, and this is often reflected in poor quality products and services. Work systems based on fragmented task allocations can also be inflexible. The skills base is low, and the movement of staff from operation to operation can be inhibited by demarcation rules and payment differentials. The introduction of new technology, new procedures, new systems, and new products and services can therefore also be slow and time-consuming.

Despite the apparent simplicity of a work system based on task fragmentation, complex support systems are usually required for work measurement, for labour allocation and production scheduling, and to maintain the payment system for different piece rates, which, in production control and wages sections, can create significant overheads. Simplicity and low cost on the shop-floor can thus be offset by complexity and high overhead costs in support functions.

This is why you may consider experimenting with other techniques based on multi-skilled team approaches which have been used in several enterprises – with some astonishing results.

Job rotation, job enlargement
and job enrichment

Early ideas, which can still be useful in some circumstances, aimed at giving people more to do than just the 30-second cycle task. The main forms of these techniques were job rotation, job enlargement and job enrichment (see figures 5.1-3).

Figure 5.1   Job rotation: People move at predetermined intervals from one task to another

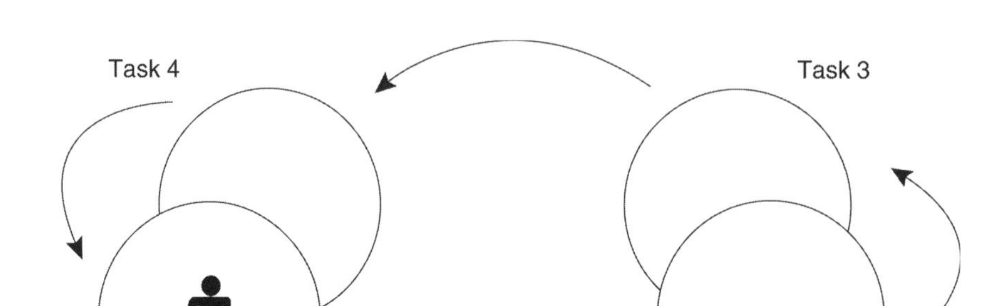

Source: European Trade Union Institute, 1991.

Figure 5.2   Job enlargement: Each person is given more tasks to perform, typically at the same level of work, so this is also known as horizontal enlargement

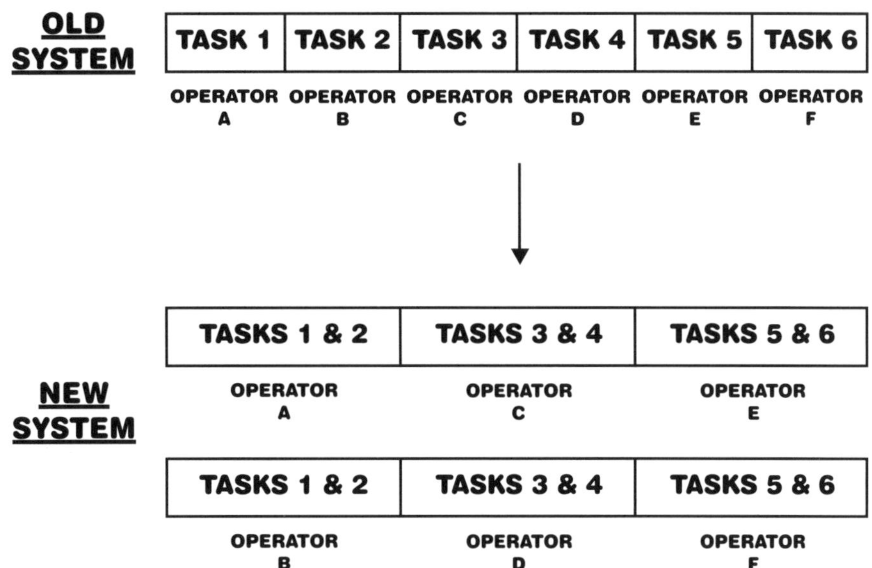

Job rotation and job enlargement have been widely applied, and can help to counter some of the limitations of the traditional approach by broadening the variety in the work experience of the individual employee. These techniques have been shown to improve productivity and job satisfaction.

However, it should be noted that these techniques apply in situations where the core work design technique is still traditional management and task fragmentation. Job rotation typically involves moving the individual worker from one fragmented and repetitive task to another at predetermined intervals. The employee might thus perform several boring tasks during a working day instead of one. Similarly, job enlargement involves combining fragmented tasks into the one job. The employee might thus perform a collection of boring tasks at the same time.

The techniques of job rotation and job enlargement involve limited changes to the nature and experience of work. The impact of these techniques on productivity and job satisfaction can be similarly restricted. These approaches may prove useful, however, in at least two situations:

1.   There may be work settings where the more sophisticated solutions described later in this chapter cannot be applied, perhaps because of the nature of the work, the work setting, or the skills level of the workforce.

2.   These techniques may be applied in the initial stages of a long-term programme of work system improvement which may ultimately involve more significant skills growth and flexible team working. An example of successful job rotation is given in the following case-study from the Philippines.

CASE-STUDY 5A

Introduction of job rotation in a factory producing plastic goods in the Philippines

In a small factory producing plastic goods, a particular job was disliked by the workers. The job was to remove excess plastic from the open end of plastic bottles after they had been made using injection moulds. The task involved taking the bottle and putting its mouth around a small grinder while turning the bottle round. When finished, the bottle was dropped into a large plastic bag. The task was uninteresting, although necessary. Only a few seconds were needed for the polishing of a bottle mouth. The work was done while standing, as some force was necessary to turn the bottle properly. Compared with the other jobs in the factory, such as operating the injection-moulding machines, labelling, inspecting, packing or preparation of materials, the job seemed too simple and boring. The person doing the job was also isolated from other workers. Productivity was low.

On hearing that everybody hated this job, the manager adopted a rotation scheme including this and some other jobs. In the scheme, a group of several workers took turns and worked at one of these jobs for only one day at a time. After polishing bottle mouths, each group member could do other jobs on the subsequent days. The other jobs were done while seated with other workers around a large work table, which workers preferred.

No direct costs were incurred in adopting the rotation scheme. The scheme was welcomed by the workers as they no longer risked long turns on disagreeable jobs. Relations between the workers also improved. The new scheme apparently had favourable effects on the workers' morale.

Source: Kogi, Phoon and Thurman, 1988.

Job enrichment involves the application of "vertical loading" factors typically incorporating some tasks previous carried out by inspectors or supervisors, so this is also known as *vertical enlargement* (see figure 5.3).

The enrichment of jobs to improve skills growth, motivation and job performance, relies on the application of *vertical job-loading principles*. These involve removing some management controls on employees, increasing individual accountability for work, giving employees complete or natural units of work and additional authority, freedom or discretion, providing feedback on performance directly to the individual rather than through supervisors, introducing new and progressively more challenging tasks, and assigning specialized tasks at which employees can become experts.

Unlike job rotation and job enlargement, applications of job enrichment have not been primarily restricted to low-grade manual work. Several white-collar applications, affecting administrative and clerical work, have also been reported.

Job enrichment has a number of advantages. It is relatively easy to understand. In almost every work setting, it is possible to find some combination of vertical job-loading factors to apply. The evidence suggests that it can quickly produce quantifiable benefits. The concepts are straightforward, and individuals and work groups can be invited to offer their own suggestions concerning how vertical loading can be applied to their work. Because increased motivation and performance depend heavily on quality of work experience and

Figure 5.3   Job enrichment

OLD SYSTEM (FRAGMENTED JOBS)

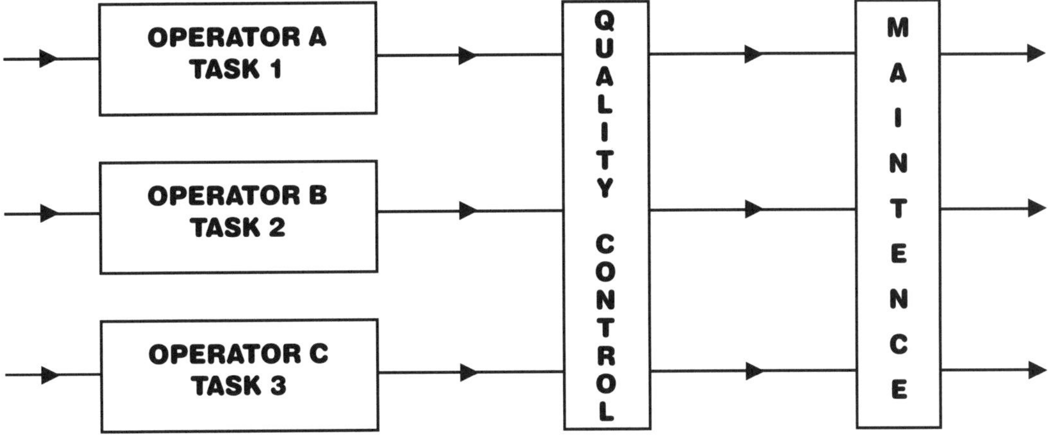

NEW SYSTEM (ENRICHED JOBS)

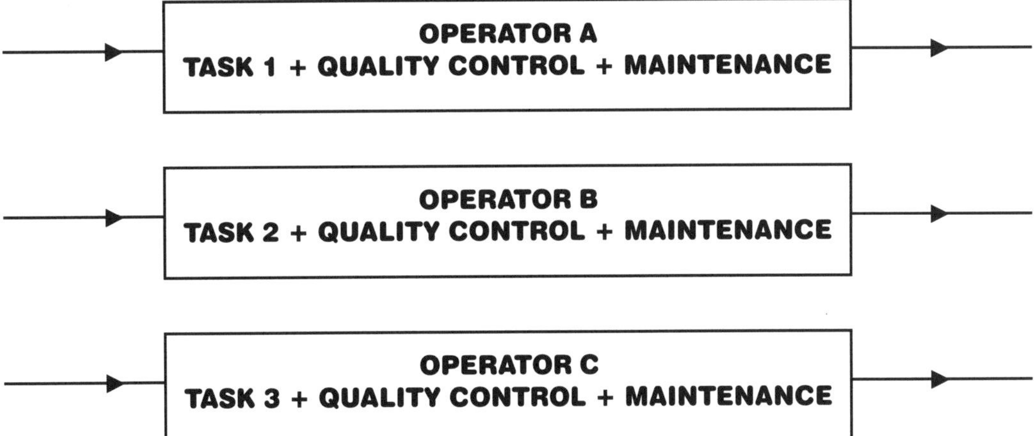

not on financial reward alone (in theory), the approach suggests ways of improving job satisfaction and productivity without additional expense.

There are, however, a number of disadvantages. There are some settings in which vertical loading is difficult, say with self-contained and isolated jobs dependent on single items of expensive equipment or processes. Vertical loading can affect supervisory roles and cause resentment and frustration among first-line and other levels of management. Resentment can also arise if management are seen to be demanding more responsibility and effort without financial compensation. The technique is ultimately limited, in its implications and benefits, by the fact that organizational structures and the traditional role of management – with the possible exception of middle management – are not affected. The limited "one-off" change can thus produce short-term, unsustained benefits. Where performance problems are created or reinforced by wider organizational and management issues, job enrichment alone may produce only limited returns.

Benefits and limitations of job enrichment

Benefits	Limitations
Easy to understand	Can affect supervision adversely
Easy to apply	Not always suitable
Widely applicable	Can cause resentment
Possibility of rapid results	Short-term benefits
Low cost	Limited outcome
Organization structures left intact	Organizational problems not addressed

Job enrichment has been the centre of a recent collective agreement on changing the organization of work in the Pechiney group of companies, described in case-study 5B.

A simple checklist can be used to help to determine whether it is worthwhile considering the application of job rotation, job enlargement or job enrichment in a given organizational setting.

Define the problem: poor performance, low morale, low levels of commitment to the organization, lack of attention to detail and quality.

Identify the symptoms: poor timekeeping, carelessness, sabotage, accidents, high absenteeism, high labour turnover, low morale.

Possible causes: task fragmentation, short duration work cycles, repetitive tasks, monotony, boredom, no skills growth.

Possible solutions: job rotation, enlargement or enrichment, depending on skills and employee expectations.

Implementation: explain the techniques and their implications, and then invite individuals and groups to suggest possible solutions.

Teamworking

Why teamworking? Because the successful operation of any business requires that people cooperate to a common end. It is possible to imagine a system, say a manufacturing plant, where everyone had enriched jobs but no recognition that they had to work together, that the person producing the lid had to make enough to match the number of pots produced by his or her neighbour. Why not employ a manager to ensure this? Far better, surely, to put the lid and the pot producers together, give them a specification, a schedule, and any necessary resources, and let them work together to achieve the required output.

By increasing responsibility, authority and cooperation, and by allowing for higher levels of flexibility, the introduction of work teams may lead to important improvements in production and in the quality of work, as well as in job satisfaction and motivation. You have created circumstances whereby people can get on with doing the job, with the authority to overcome any obstacles and the skills to use that authority. This ability to achieve is itself both satisfying and motivating. However, the development of effective teams is a delicate exercise. Teamworking requires that every teamworker:

CASE-STUDY 5B

Job enrichment at Pechiney group – France

At the end of 1990, the Pechiney group employed 30,770 workers in France, with a further 40,000 in subsidiaries elsewhere in the world (mainly Europe and North America). Its operations are mainly in the areas of packaging materials, aluminium and industrial components. On 13 November 1991, French group management signed an important collective agreement with four trade unions (CFDT, CFTC, FO and CGC) on "the implementation of changes in work organization".

The key idea underlying the new agreement is that the increased competitiveness of the group's companies and the improvement of their employees' working conditions are two objectives which are closely linked. They should therefore be the object of an integrated approach within the framework of initiatives to change the way in which work is organized. The simultaneous pursuit of these two objectives means that the skills of employees must be managed in a forward-looking way, thus guaranteeing that the new forms of organization function smoothly and, at the same time, allowing the largest possible number of employees to adapt themselves to the new systems and to benefit from them.

This means enriching the content of work, notably by abandoning the notion of jobs which consist only of the carrying out of individual and related tasks, and creating instead fuller jobs, allowing employees both to make use of wide technical skills and to add skills in other areas. For managerial staff, organizational changes should allow them to extend their area of competence to include matters such as personnel management, the technical management of the unit, ordering and so on.

The introduction of this forward-looking skills management will, according to the agreement, require:

■ an analysis of existing jobs and functions, identifying the skills of all types being used;

■ an analysis of future requirements and of the skills needed to meet them;

■ an assessment of the gaps between future needs and current skills, and a study of the training and career paths of existing employees aimed at evaluating their capacity for change; and

■ the elaboration of measures to reduce the gaps identified.

The agreement identifies training as the key means by which the economic aims of changing work organization are to be reconciled with the expectations of employees in terms of the development and recognition of their skills. The aim of training should be twofold. First, it should provide the enterprise or establishment with the skills required by the new system of organization. The training should thus bridge the gaps between available skills and resources and the skills needed for the implementation of the new system. Second, it should allow employees to acquire new skills and thus to progress within their occupation or career path. The training should help to prepare for the future, giving priority to employees already in place and benefiting those employees.

Source: Industrial Relations Services, 1992.

- understands his or her role and duties;

- has the necessary authority;

- takes responsibility;

- develops his or her competence and skills; and

- actively cooperates with other workers to the achievement of common goals.

But, given that we have achieved this state, we have still to get from where we are now, to a situation where the teams are operating. This journey is as important as arriving at the target, as we learn about the new situation by working together to get these changes.

Setting up teamworking

First, let us look at how to operate in setting up teamwork. While senior management will set policy, the decisions and changes in the organization are best done by those closest to the work itself, for they know best the effects of, and the possibilities for, changes. These working teams, coordinated by a managing team which includes working team members, should be as small as practicable and composed of people from the managers, engineers, service personnel and workers involved in the sector under consideration. This, together with the skills of the team, is set out below.

A combination of the following skills will greatly add to the output of the team:

- *technical expertise*: on teams seeking to reorganize, design or manage projects, it is essential for members to be able to make informed and critical choices on the hardware and software aspects of change. Technical expertise does not have to be lodged with every member but must be represented;

- *organizational expertise*: the acceptance and impact of decision-making by the team will be facilitated if they have a good understanding of how the organization works, how change is accommodated and what it will mean in terms of changing the relative power balance or status of groups inside the enterprise;

- *interpersonal skills*: often ignored, this aspect of membership skills is vital to the creation of good dynamics. The ability to listen as well as suggest, to support rather than confront, and to act constructively in a group can make the vital difference in terms of team effectiveness.

A team's effectiveness should be "balanced" with a good blend of skills and personalities, neither too homogenous nor too heterogeneous.

The teams will need to have a systematic way of operating. A standard project management approach, as outlined below, is appropriate.

A standard project management approach

1. Develop strategy

2. Confirm top-level support

3. Use a project management approach:
   - identify tasks
   - assign responsibilities
   - agree on deadlines
   - initiate actions
   - monitor
   - act on problems
   - close down

4. Communicate results.

These points suggest an iterative process be adopted through which team objectives emerge. It does also propose firm and grounded information-gathering procedures on which to judge results and apply adjustments.

Solving the problems

There will be many problems to be worked through – technical, organizational and social – and an understanding of diagnosis and problem-solving will be beneficial. The steps in diagnosis and solution are listed below.

Diagnosis

1. Define the problem in one short, simple sentence: this precision helps to improve focus and aids understanding of the problem.

2. Find at least two other ways to define the problem: this can also be useful in improving understanding of the problem by introducing other perspectives.

3. Identify the symptoms of the problem: leave the causes aside at this stage.

4. Identify all the possible causes of the problem: it helps to be comprehensive at this stage, just in case issues have been overlooked.

5. Identify the most likely cause or causes: here precision and focus are restored.

Problem-solving

1. Brainstorm possible solutions to the problem: it is necessary at this stage to take a creative approach which may produce novel and effective solutions that had not previously been considered.

2. Evaluate the most appropriate solution or solutions, remembering that, depending on the nature of the problem, multiple solutions may be appropriate.

3. Implement the solution or solutions, using a project management approach (see above).

This diagnosis and problem-solving structure works particularly well with groups which have access to a facilitator or instructor to help them work through the procedure systematically. However, this methodology can easily be used by local task forces or project teams working on their own. The group addressing the problem should include those who are affected by and who understand the problems being addressed. The facilitator should be able to focus the attention of the group systematically, step by step, through the structure, and prevent open-ended or unstructured discussion around the problem issues. With large groups, it is extremely useful to split into syndicates with three to five members each, with each syndicate developing its own analysis through the same problem-solving structure, for subsequent presentation to and comparison with the other syndicates once the process has been completed.

To be a team, each member has to know what is going on. Information is crucial to a well-functioning group, especially one like a design team which is operating in a context of change and uncertainty. Obviating anxieties arising from change needs the establishment of an information system which can keep the team in touch with both long- and short-term changes. It is also most important to keep all who might be affected by change in touch with progress: members of the team need the support and feedback from their "constituencies". Therefore bulletins, meetings and ongoing discussions about progress are necessary.

Information for the team will include the matters on diagnostics and feedback given below, which can be seen, among other things, as part of an agenda for team meetings.

Diagnostic information on:

- where current designs fail
- the significant variables involved
- controllable/uncontrollable variables
- organizational values underpinning production
- institutional constraints, power, status issues, etc.

Feedback information on:

- observation on modifications
- continuous performance information and up-to-date productivity indicators
- workers' reactions to change
- turnover, sickness, absence, etc.

There are other areas where the effectiveness of the team can be supported or handicapped unless the actions taken are clear and conclusive. Changes such as those described here affect more than just direct workers: restructuring people's jobs to provide, for example, more autonomy in action and decision changes management's roles at all levels. Hence the redefinition of these roles is necessary, so that people know where they are and what is needed from them. This changes the culture of the organization, which also alters how performance is monitored. Last, but by no means least, training will make a major contribution to success. Different jobs and different roles will require new knowledge and skills, and these must be provided; learning on the job can provide some, but where this is insufficient, slow or uncertain, then common sense requires that formal training is needed.

Middle management

With respect to management, organizational hierarchies have in some companies been flattened, with middle and junior managers assuming other internal consulting and advisory roles within the organization in place of directing or line functions.

For this to happen, the company culture and career systems have to allow and encourage managers to step out of conventional positions and career ladders into other areas and still feel secure. Table 5.1 summarizes the differences.

Table 5.1 Management changes with teamworking

Traditional context	Traditional functions	
Stable environment	Plan	
	Organize	
Simple technology	Coordinate	
	Motivate	
Standard output	Control	
Emphasis on growth, cost reduction, labour control		
New context	**New functions**	
	Create the vision	
Turbulent	External focus	Internal focus
Sophisticated technology	Scan environment, consult, find new business, support	
Changing outputs	Manage vendors, improve logistics	
Emphasis on flexible response to change, quality, employee commitment		

First-line supervision

With respect to the role of the supervisor, some organizations have eradicated this function altogether, while some have transformed the function from a directive to a facilitative one – from "policeman to coach". First-line supervisors in this approach lose their traditional problem-solving, fire-fighting, progress-chasing, directive role and become counsellors, consultants, team facilitators and trainers. The advantages of this strategy arise from the business improvement contributions of supervision, and the potential for skills growth and rapid problem-solving on the shop-floor.

This transition has a number of key dimensions, which are summarized in table 5.2.

Table 5.2   The changing role of first-level supervision

Traditional	*New*
Span of control = 10+	**Span of control = 50 to 75**
Scheduler of work	**Coach and sounding board for self-managing team; leader/coordinator working on training to emphasize skill development**
Rule enforcer	**Facilitator, getting experts to help teams as needed**
Lots of planning	**Lots of wandering around**
Focused up and down the structure	**Focused horizontally, working with other functions to speed action**
Transmitting management	**Selling team ideas and needs up, company needs down**
Providing new ideas	**Helping teams to develop their own ideas and providing ideas for cross-functional systems development**

Source: Peters, 1987.

Training

From the information above, it is clear that all levels of the organization will need some training to help with the changes. When we began this section on teamworking, we gave a list of five requirements for a teamworker. It will be recognized that some of these will arise from new skills, and others from the changes to both jobs and organization. To institute changes to jobs and organization will also need new information and skills, some of which will have to come from training.

Hence the training needs can be extensive, they can be time-consuming, timing the training itself can be problematic, and the total training package can be expensive. In general terms, training needs are likely to fall into four critical categories:

- new concepts awareness;
- facilitative supervision skills;
- multi-skilling through cross training; and
- training in group problem-solving and decision-making.

First there has to be a widespread understanding, among management and employees at all levels, of the nature of teamworking, assuming that the organization and its members have no prior experience of this approach. For some organizations, this represents a radical departure from custom and tradition. Training and development programmes can be important in beginning the processes of attitude and behaviour change.

Second, first-line supervisors and team leaders require skills training in their new facilitative role, with respect to self-managing groups of employees. First-line supervisors can find this transition difficult because, for them as individuals, it can mean a radical change in customary, and comfortable, behaviours. This can be particularly acute with respect to perceived loss of status. It can also be difficult for experienced supervisors, skilled and experienced in solving operational problems, to stand back and let their subordinates solve those problems, perhaps making mistakes in the learning process. Training here serves the substantive function of identifying and transmitting new skills and behaviours. Training also serves the symbolic function of signalling the change in role for the supervisor, as this usually involves absence from the organization for a period and a change in job title. Titles vary, but typical examples include shift coordinator, team leader, group facilitator, and other combinations of these terms.

Third, team members require cross training in the relevant skills range if they are to become truly multi-functional and multi-skilled. In a changing organizational environment, this is likely to require continuous retraining too. Digital (see case-study 5G) had their equipment maintenance engineers train shop-floor employees in basic machine maintenance and repair routines, freeing the technicians for other logistical work. Training can be off-the-job in this way, but can also – as in Digital – be carried out within the teams. Experienced team members can be expected to train new and less-experienced members in a range of product, process and equipment skills, but this needs monitoring if poor practices are not to be entrenched.

Finally, it is rarely realistic to allocate employees to teams and then expect them to function immediately as a team. Training will almost always be required in group processes such as team building and team working methods. This is particularly important with respect to team problem-analysis and problem-solving techniques, and also with respect to group decision-making. Where teams are expected to operate without direct supervision, training in basic communication skills, presentation skills, and meeting chair skills can also be useful. These training requirements can easily be overlooked, with an understandable emphasis on job-specific skills. However, if teamworking is to be fully effective, these skills areas must also be addressed.

There are, however, some problems with this transition. New selection criteria for supervisors may be required, along with retraining, and revised payment and career systems. As already mentioned, some supervisory staff find this transition extremely difficult. Further reasons for change have to be supported by extensive communication so that all those concerned are clear about the new supervisory role and that this is introduced and adopted consistently. Off-the-job group working and role-playing can help in giving people a feel for their new roles.

Organizational support

Finally, we have the factor of organizational support, the glue which binds people together and assists them towards a common purpose. Teams do not work in isolation from other parts of the enterprise nor from the culture and general strategies that they employ. The context is important in giving or withholding organizational supports. Here we wish to mention three out of many that could be discussed.

Rewards and recognition. For many enterprises, the drive towards decentralization and teamwork has proceeded alongside the movement away from individualized pay and individual bonuses. Rewards related to team performance, whether financial, status or recognition, have proved to be a most appropriate support to effective teamwork.

With respect to financial rewards systems, two trends have become evident. These concern the development of skills-based payment systems. Pay is then based on the individual's skills profile. Job skills are assessed through a formal test-based certification process, while other skills are assessed by colleagues on the team. This represents a fundamental departure from traditional job evaluation methods; it is the skills in the person that are assessed and that attract reward, not the skill level of the job the individual happens to be doing at any particular time. This approach has the effect of increasing employee motivation to develop business-relevant skills.

Gain, or profit-sharing, is another developing form of financial reward. It can take many forms and have many different bases. It can be related to costs, productivity, value added, turnover or profit. The most appropriate method is dependent entirely on local circumstances.

If there is one single, main principle of effective gain-sharing, it is that the rewards must be seen to be closely linked to employee efforts and contributions. Where that link is obscure or does not exist, the gain-sharing scheme will have little or no impact on employee behaviour, attitudes or performance because the rewards are seen not to depend on behaviour or attitudes in a meaningful way.

Organizational culture. The collective norms and beliefs of an organization can stifle or encourage experimentation, innovation and creative human resource usage. It is the case that most examples of teamwork have been part of wider changes in management outlooks and systematic organizational development. Cultures that are open, participative, that emphasize quality, performance and innovation, do appear to have less difficulty in applying teamwork and the more advanced variants of it.

Monitoring systems. This aspect is closely related to the organization's ability to provide a good information system that can allow for close monitoring of performance via regular reviews.

Regular review can improve team performance by:

- ensuring that adequate effort is directed towards planning;

- improving decision-making;

- increasing support, trust, openness and honesty;

- clarifying objectives;

- identifying development needs and opportunities;

- increasing the effectiveness of team leadership;
- making meetings more productive and more enjoyable;
- decreasing the number of emergencies and crises;
- increasing involvement and commitment.

It is simply a question of learning how to function more effectively in the future by looking at the way the team is operating in the present.

To assist you, box 5.1 below suggests a simple procedure to help in clarifying how monitoring will take place. If you are working as a member of a project group, work through these questions first on your own, with a view to comparing and discussing notes later with your colleagues. Then draw up an agreed monitoring plan, and make sure that someone is responsible for implementing it, and reporting back to the project team, or to the appropriate manager.

Different working teams at work

Various types of teamworking, from the more simple to the more complex, have been developed within enterprises to meet production needs efficiently, and you may also wish to consider introducing or expanding team working in your enterprise based on the following case-studies (case-studies 5C and 5D).

Box 5.1   Six questions to help in defining monitoring requirements

1.   **Why do you want to monitor progress?**
      For example: how do you think this will help you?
                     are there any particular aspects you want to check?

2.   **What measures will you monitor?**
      For example: cost reduction?
                     productivity improvements?
                     faster information flows?

3.   **Can you establish specific targets for each of those measures?**
      For example: 10 per cent less scrap?
                     24 per cent more output per head?
                     daily instead of monthly reports?

4.   **What are the review intervals and dates?**
      For example: monthly review?
                     what will be achieved after six months?

5.   **How will monitoring information be collected?**
      For example: from existing data flows?
                     from a special review procedure?

6.   **Who will be responsible for conducting the review?**
      For example: the project leader?
                     managers of user departments?

Source: Boddy and Buchanan, 1987.

CASE-STUDY 5C

Teamworking at Express Tanzania Ltd.

This company came into existence in the early 1960s, being one of the Lonhro (a British consortium) group of companies. Its key operations are in the clearing and transportation of goods. Over the years, the company grew steadily and acquired a good name as a reliable international clearing and forwarding enterprise. By 1977, it employed some 180 people with a local management team and no foreigners in it. Administratively it was still under foreign guidance from the Lonhro headquarters in London. Apart from administration, there are four main sections: maintenance, packing, carpentry and heavy-lift sections, all located in the industrial area of Dar es Salaam.

The project started in mid-1977 in the vehicle maintenance department and continued later in the packing, heavy-lift and carpentry sections.

The maintenance section

There are four main tasks performed in this section: mechanical repairs, electric wiring, panel beating and tyre repairs. The section repairs all the company's motor vehicles as well as those which are damaged in road accidents where company vehicles are involved. Twelve people are permanently employed in this section.

The traditional work organization was based on rigid specialization. Only four out of the 12 were skilled mechanics, one for each type of task mentioned above. They did the job which needed their special skills, while the rest were either assisting or they were tyre-men, the least popular job. This distribution of work led to much waiting and much rushing to get vehicles back on the road after repairs. Quite a few jobs had to be done by outside shops. The skilled workers were reluctant to pass their skills on to those semi- or unskilled. An increasing gap, professionally and socially, developed between the mechanics and other workers.

When the project was launched, these problems were discussed openly with all involved. Brief explanations were given to all the workers of the possibility of trying alternative forms of work organization. The local trade union was actively involved at this stage as well as the Workers' Education Division. The response among employees was positive. However, since the workers were on widely different levels of skill and pay, it was difficult to go straight into work reorganization.

The company therefore decided first to make sure that all the workers in this section acquired some basic skills in the four main tasks. This was arranged by having each of the four skilled workers accepted to train three unskilled workers for a period of three months. After the three months the teams of three were rotated to be trained by a different skilled mechanic. Regular evaluation of the effectiveness of the training was done. In 12 months, an overall evaluation was done. All except one of the semi- and unskilled workers grasped the basic skills. In addition to the rotational training system, those who were interested were given the opportunity to take part-time evening studies at the nearby Dar es Salaam Vocational Training College. Here they learned more theory and were able to acquire government-recognized trade test certificates in the various fields of study. All the costs for such studies were paid by the company.

Case-study 5C (cont.)

When the major training programme was over, teams were formed and the jobs were allocated as group tasks. To avoid monotony, the teams also rotated in performing their tasks. This helped to broaden the training and experience of the various types of tasks.

By the time the project entered its final phase after two years, the results of the reorganization started to be quite visible to workers, management and the outside advisers. The results were as follows:

(i) More cooperation and teamwork while performing the tasks. Entering the workplace, one would immediately notice people discussing how to do the work together and, more important, actually sharing tasks and responsibilities. Very few jobs were delayed and if they were, the reason was mainly lack of spare parts.

(ii) The company no longer sent their vehicles for any type of repair to other garages since now all the repairs could be done at the workshop of the company. This had substantially reduced the costs of repairs to be incurred for paying other garages. No figures could be given, but company management judged this alone to be important enough to continue the project, or rather to accept the results as company policy.

(iii) Increased learning on the job. This was clearly seen in the way jobs were reduced.

(iv) Complaints about future prospects and concerning opportunity of expanding the skills were reduced. Very little friction was seen between workers on different skill levels.

(v) Upgrading of workers. Two of the mechanics who had no formal grade in this trade attended the Vocation Training Centre and, by September 1979, passed the exams and acquired Grade III Trade Test Certificates as mechanics. Another noticeable effect of project implementation is the overall upgrading in the form of increase in salaries by all the workers in this section. On average, their salaries increased from T.Sh.483 in 1977 to T.Sh.573 in 1979. (This occurred in a period when no general increase had been introduced by national agreement or government regulation.)

Source: ILO, 1981.

CASE-STUDY 5D

Teamworking at Opel

In 1991, an agreement was reached between the management and works council of Opel to introduce teamworking within the company's plants in Germany. The agreement included the following provisions.

Aims

Increased flexibility, greater job satisfaction, continuous improvement and productivity increases.

Functions

Teams are to consist of eight to 15 members and to be responsible for:

- the internal allocation of work;
- scheduling breaks;
- shift transfers;
- team meetings;
- coordinating holidays;
- improving health and safety;
- meeting qualitative and quantitative production targets;
- maintaining the lines of communication with the supervisor;
- induction of new members;
- keeping the working area clean and tidy;
- encouraging "creative, innovative and independent thinking" among team members;
- fostering their participation as regards the content and organization of work and the working environment.

It is envisaged that individual workers should be able to work flexibly and undertake different tasks within the group where this has the agreement of the team and of the supervisor.

Team meetings

Team meetings are to be held on a weekly basis. Where possible, such meetings are to take place within normal working hours. Should this not be possible due to production or other imperatives, the meetings will take place outside working hours and be subject to overtime payments. Groups determine the subject of these meetings themselves and are free to invite members of the works council, technical experts or members of management to help clear up specific questions. Supervisors, works councillors and technical departments may also introduce subjects for the agenda of group meetings and participate in them.

Case-study 5D (cont.)

Team leaders

Each team will be represented by a spokesperson or team leader (*Gruppensprecher*), who is elected by the team members. Team leaders do not have any powers of direction or discipline. Their main responsibilities as defined by the agreement include motivating the team; reconciling differences of opinion within it; ensuring that group targets are met; representing the views of the group towards third parties; and summoning technical assistance, if necessary. Group leaders are elected for a six-month period in the first instance and for a further year if they are re-elected.

Supervision and training

Teamworking has fundamental implications for the role of supervisors. The agreement foresees that their responsibilities will shift from a managerial emphasis towards an advisory and supportive one. As such, their main duties will consist of proposing and agreeing targets; providing assistance where the group encounters difficulties in meeting its objectives; providing support in relation to problem-solving; and maintaining communications and coordination beyond the level of the group.

Before teamworking is introduced, all those involved will undergo preparation as regards their future tasks. Team leaders, their deputies and supervisors will be provided with additional training in areas such as leadership skills and conflict resolution. Specific training plans are to be elaborated in cooperation with the works council and will require its approval.

Pay structure

A new rationalized pay structure is introduced. This reduces the number of wage groups from 42 to ten, which now include all bonuses and piece rates. As a result employees will receive average wage rises of 3 per cent, in addition to the recently agreed industry-wide increases of 6.7 per cent.

Source: Industrial Relations Services, 1991.

Autonomous working groups

We said earlier that the journey was as important as the destination, and in the journey through this chapter we hope that you see why, at the end, we are presenting this material on autonomous working groups.

The trend through the chapter has been from minimal change, through job enlargement with positive but minimal benefit, to major alterations in how work is done, with major benefits. An autonomous working group is a working team with the full power to focus its attention on its objectives, with the structure of the organization there to support its activities in this direction. Hence all that has gone before about introducing teamwork is relevant, but it is useful now to focus our attention on what can be seen as a mature work organization for the future. Note, incidentally, that much of what has been said applies to "teleworking", that procedure increasingly used in the information industries where people work on their computers remote from the central office. Here the individual is, in many ways, a "team of one", and needs similar organization, support and training as does a team.

The objective of a working group is to produce the required output, be it an insurance policy, an income tax assessment, a car seat frame, or a serviced and repaired washing machine, at the required quality and price at the required time. The purpose of the surrounding organization, the firm, is to back up this objective with all necessary resources, including information and policies. These requirements illustrate why autonomous groups exist, and define the limits of the self-regulating and self-managing of these groups. The limit is a boundary within which they can take all necessary decision and actions to achieve their objective.

A more formal definition might be as follows: an autonomous work group is a team of people who have been allocated a bounded set of work activities and who are given total (autonomous working groups) or partial (semi-autonomous working groups) discretion concerning how those activities are to be carried out and shared between them. This discretion or flexibility is also called responsible autonomy. Figure 5.4 shows how previous areas of supervision and management are incorporated into the functions of autonomous working groups.

Why an autonomous working group?

Autonomous groups have come about for several reasons, some of which have been outlined earlier. As it is important to be clear why we pursue changes in this direction, they will be set out again here.

Increased commitment

Modern business and industry require that people should work effectively and with a commitment to achieving objectives of quality, output, timing, etc. If people are to involve themselves, then they must have enough job content to involve themselves in, enough variety to maintain interest, enough authority to take any necessary action to get on with their work, and enough skill and knowledge to do what is needed well and correctly. Motivation and interest arise from skills used freely, at the decision of the doer, to successfully achieve the objectives, and the consequence of such activity is to reduce the effects of alienation from work, such as boredom, absenteeism, turnover and sickness absence.

**Figure 5.4 The autonomous group
and the devolution of responsibilities**

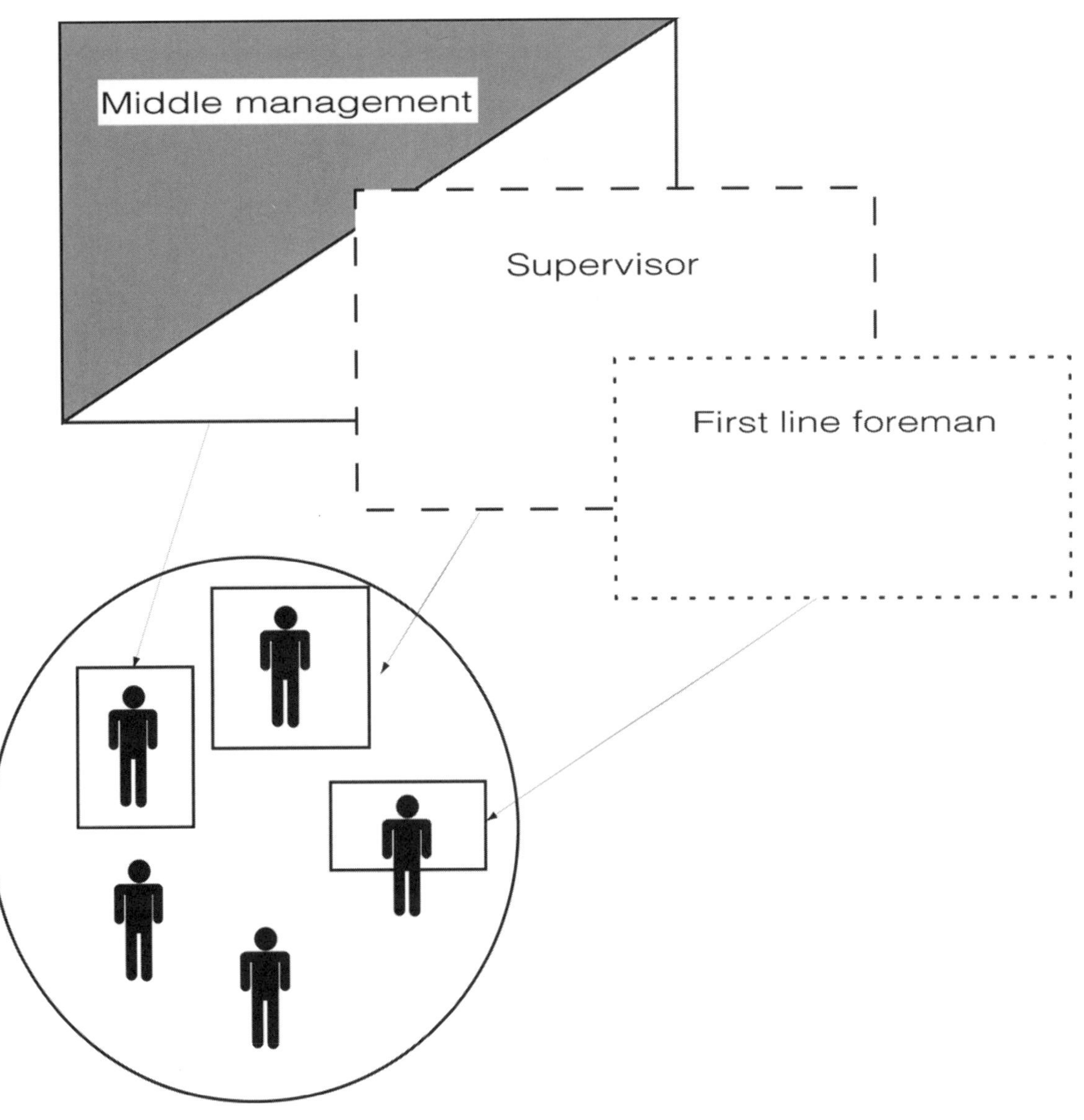

Source: European Trade Union Institute, 1991.

Improved quality

An increasingly sophisticated global market-place has forced many organizations to pay much more attention to the quality of the service or product that they sell. This has led many service organizations to see the customer as part of the team, the end point of the work flow, and to develop customer-care programmes, to train staff who come into regular contact with customers to be more sensitive to customers' needs, and to be more responsive to their demands. This same trend has led many manufacturing organizations to develop total quality management programmes to influence staff at all levels to pay more attention to getting it right the first time, to zero defects, and to reduce time to market. A team with control over its product can do this, an individual on a line has much less influence.

Flexibility and responsiveness to change

The need to design organizations and work systems that can respond effectively and flexibly to change is pivotal in modern enterprises. It has been clear for some time that jobs and organizational structures designed according to traditional principles of task fragmentation are not effective in coping with rapid change. In order to adapt work organization to meet diverse and volatile market demands quickly and flexibly, rigid specialization based on an articulated job classification scheme may not be conducive to the efficient utilization of a workforce. It may become necessary to assign diverse tasks to workers flexibly, responding to evolving circumstances and the rapid change of tasks, while maintaining coordination in output is particularly suited to teamworking.

Teamworking may also be introduced in the wake of new technology. New machinery or equipment may increase the production workers' freedom of movement. Rather than being tied to an individual workstation, production workers may assume a more general responsibility for the production process. This requires a deeper understanding of the process and at least some problem-solving abilities.

Responding to local shocks

It may also become necessary to cope with local shocks quickly (such as the malfunction of machines and absenteeism) without calling in specialized help (such as by engineers, service staff and relief staff) from outside the shop. But doing so may require more versatile skills of workers and team-oriented effort.

Direct information and decision-making

The centralization of information and decision-making may be subject to disturbance in the process of communications and time-lag from the perception of change to the implementation of operational response. Also, if intermediate product flow is controlled centrally, valuable on-the-spot information available at interfaces of constituent units, such as on the quality of intermediate products and emergent events affecting timing of delivery, etc., may remain unused.

Experimenting with autonomous working groups

The three case-studies which follow (5E, 5F and 5G) demonstrate that there is no one best way to implement flexible teamwork. The approach has to be tailored to suit local needs and circumstances. The language of these applications also varies from setting to setting, although the underlying concepts and the nature of the applications are similar. Autonomous groups, integrated working, high performance systems – are all based on the same core ideas and all have similar implications for employees, management, organization and employment policies.

CASE-STUDY 5E

Semi-autonomous working group at Bharat Heavy Electricals Ltd. in India

Bharat Heavy Electricals is one of the largest public enterprises in India, with four major manufacturing plants in existence and two under construction. Some of the existing plants, such as those at Tirhchirapalli and Hyderabad, have been undergoing expansion with a view to covering a wider range of products. The enterprise employs over 56,000 people in manufacturing, marketing, research and development, servicing and corporate management. Some of the output is exported.

The Hardway unit, which employs over 9,000 people, is mainly engaged in the manufacture of heavy electrical equipment such as steam and hydraulic turbines, generators and related equipment. Block V in the Hardway plant, where 25 workers were engaged in fabrication of the upper part of a condenser unit, was selected for the pilot experiment in view of its compact character, reasonable layout and the positive attitude of the manager and the shop-floor trade union leaders.

The workers covered by the pilot experiment participated in a series of meetings with both internal and external consultants, and agreed that the work should be redesigned. The total complement of 25 workers in Block V was made up of nine fitters, three fettlers, three welders, two gas-cutters, one crane operator, two riggers, two helpers and three workmen involved in materials supplies.

After deliberations, the task force, with the concurrence of the employees concerned, evolved a work system in which the direct production group would consist of one welder, three fitters and one fettler. The group should take charge of the complete task, and its members would gradually take up one another's job by undergoing on-the-job training assisted by the supervisor, the industrial engineer and their fellow workers. Cross-skilling would also be introduced between the crane operator and the riggers. It was decided that the gas-cutters and helpers, on the one hand, and the materials supplies group, on the other, would be integrated into the new work system at a later stage. As the system came into operation, a number of advantages and drawbacks emerged. On the positive side, a welder now started working as a fitter and if he did not know the art of reading elementary drawings, which was a necessity for a fitter's job, arrangements were made by the management to provide on-the-job training. Similarly, training on the other trades of their work groups was given to fitters, fettlers, gas-cutters and others.

With more experience and confidence, the workers cooperating with the task force brought about a second redesign of their work organization in September 1975. It was decided that the workforce should be distributed in two shifts in the following manner:

Shift I	Shift 2
5 fitters	4 fitters
5 welders	4 welders
1 gas-cutter	1 gas-cutter
1 fettler	1 fettler

In addition, there would be crane operators in both the shifts. Each shift group became integrated and self-contained. It was decided that one shift would fabricate the right side of the upper part of the condenser unit; the other shift would do the same with the left side. As and when a pair of sides were completed, the two groups would then weld them together in order to complete the upper part of the condenser.

.../...

Case study 5E (cont.)

Progressively the new work organization were extended to other blocks within the unit. With the introduction of the new form of work organization, there was an improvement in productivity, although there were also occasional setbacks. Progressively the old pattern of one-man, one-function was superseded by the acquisition of multiple skills and the development of a group system of working, with internal monitoring of group norms, internal control of work flow and work allocation, and identification with the product and its quality. Particularly significant from the management's point of view was the gradual drop in personal idle time (e.g. time spent loitering about at the workplace, or spent outside the workplace without justifiable reasons). The commitment of the workers, the supervisor and the manager was distinctly apparent.

The reorganization of work led to enlargement of the supervisory role in the form of liaison between the input and output departments and service units, and involvement in central planning. This became possible because the work groups assumed substantial control over the production process, including routine inspection and maintenance activities, in addition to implementation of rules of discipline. All this experience encouraged the management and the task forces to set up separate project teams to deliberate on and submit reports on the following three themes:

(a)   multi-skill training, its role in employee satisfaction and higher productivity; encouraging the motivating factors for workers to acquire skills in different grades;

(b)   the changing role of supervisors, particularly in respect of coordination, planning and training; and

(c)   diffusion of the scheme of work redesign to white-collar staff working in such areas as the personnel, finance and medical departments.

Source: ILO, 1979.

CASE-STUDY 5F

Autonomous working groups at SKF-D3 plant in Gothenburg, Sweden

The SKF group consists today of approximately 150 companies with some 80 factories in 17 countries. The number of employees is approximately 45,000. SKF is the world's leading manufacturer of ball and roller bearings.

In 1984, in a general group effort of modernization, the Swedish factory within SKF Norden, known as D3, was chosen for an experiment of renewal both of the production apparatus and the work organization. In 1986, a number of concrete measures were brought into play.

The creation of development groups

These groups included almost 50 people, equal to a quarter of the whole workplace at the time, who had been appointed by the employees.

One important principle was that those individuals who could actively contribute something and who were really interested were chosen. Thus there was no assumption that unions and management should have a certain level of representation. It was rather the knowledge and experience of the person concerned that was the deciding factor. The groups included representatives of the workers, salaried employees, factory management and researchers.

To start with, four development groups were created for each of the following areas: new technology, work organization, training, and work environment.

The point of departure for the work of the groups was that no given solutions were available. The result would depend to a high degree on the work of the group participants within the four areas. Another of the conditions was that it was permissible to make mistakes and to look at different solutions on a trial-and-error basis until the correct one was found. The groups were also to be truly future-oriented and were to attempt to raise themselves above everyday problems.

Hierarchy broken down

A change was also made to the overall organization, which was altered radically. SKF Norden has been changed from a functionally oriented business to a market-oriented, commercially minded enterprise. These changes were made as a result of the new philosophy which was becoming increasingly characteristic of the company. In actual fact, the reorganization meant that the hierarchy was broken down and that the organization became "flatter".

Levels before reorganization	**Levels after reorganization**
1. Managing director	1. Business area manager
2. Production manager	2. Factory manager
3. Departmental manager	3. First-line manager
4. Section manager (highest level manager at D3, for example)	4. Process executive
5. Plant engineer	
6. Production supervisor	
7. Foreman	
8. Machine setter	
9. Operator	

.../...

Case-study 5F (cont.)

Extended, tailor-made training

The training programme's aim was that the personnel in the production groups would initially achieve 50 per cent knowledge overlap. In time, most people should be capable of dealing with all the tasks which are assigned to the group.

A special training department was set up locally within D3, which was responsible for both basic and further training. In the past, training had been run centrally. Since the induction, further training and broadening of job roles was the responsibility of D3 itself; the personnel learned much more quickly. In addition, a new training methodology was based on the idea that workers should start by "unlearning" everything they used to know. That is to say, they were made fully aware that it was no longer possible to go on working in the old way.

The creation of autonomous working groups

The next step in the process of development was to create an effective work organization with production groups that were given greater work content. The D3 factory was in the middle of this phase in the early summer of 1987. The new work groups were to be autonomous and were to be responsible for a number of new tasks in addition to production.

The following new tasks were successively handed over to the groups:

- **Planning**
- **Budgeting**
- **Measurement of productivity**
- **Requisition of components**
- **Induction and initial training**
- **Grinding, control and assembling, i.e. production**
- **Maintenance and repair work**

In practice, this meant that the group itself was to monitor and control the process and make any adjustments which might be necessary. In the past, the operator was tied to a single machine. Now he or she is responsible for the whole process within the framework of the working group. As the work each job involves is broadened, responsibility for quality, volume, delivery times and delivery reliability, as well as costs, has increased. For example, the operator can quickly act and prevent faults, which means fewer rejects. The work organization is now being introduced more widely, which means that the groups are to be given a regular, fast feedback on the results of their work, which is important for job motivation. At a later stage they will also be able to measure their own productivity.

Results

The D3 factory was something of a problem experiencing severe production disturbances as late as 1986. The development project also lost a certain amount of momentum on several occasions. But the dedicated effort gradually bore fruit. Towards the end of 1986, something began to happen which can now be seen in terms of financial results. D3 is now one of the most efficient factories within the whole of SKF Norden.

The employees at the D3 factory have taken the new impulses and ideas for change to heart. This means that the process of change will be taken further and will also be developed in future.

Source: Swedish Work Environment Fund, 1989.

CASE-STUDY 5G

High performance working groups at Digital Equipment Corporation

Another company that has been at the leading edge in introducing innovative approaches to work organization has been Digital Equipment Corporation. Their approach illustrates the increased sophistication in contemporary applications of flexible teamworking, not as an isolated technique applied to the shop-floor or office, but as a systematic, integrated package of changes also affecting organizational structure, management and employment policies.

Digital label their approach high-performance work systems design, after the pioneering work of Peter Vaill. Vaill sought to identify the characteristics of high performing systems in organization: "human systems that perform at levels of excellence far beyond those of comparable systems".

Digital Equipment first applied the high-performance work systems approach at one of their American plants, at Enfield, Connecticut. This plant made printed circuit boards, and increasing competition and product changes encouraged management to experiment with new approaches to work system design. The main goal at Enfield was flexibility: the capacity to respond quickly and effectively to a highly uncertain environment. Traditional ways to handle uncertainty had included the introduction of new procedures, changing the structure, employing more people, and tightening management controls. These strategies introduced increased overheads, created more interfaces, and increased the complexity of the organization and generated more uncertainty.

To deal with these issues, management decided to introduce more systematic career planning and development. The plant manager's review of these changes revealed the following results:

- 40 per cent reduction in product manufacturing time;

- 15 inventory turns per annum;

- only three levels of management hierarchy;

- multi-skilled operating teams;

- 38 per cent reduction in standard costs;

- equivalent output with half the people and half the space; and

- 40 per cent reduction in overheads.

This and the experience of other American corporations (Procter and Gamble, Yilog, Hewlett-Packard) encouraged Digital managers at their plant at Ayr, Scotland, to develop a similar, customized approach to work system design.

The Digital plant in Scotland employed 670 people in 1986, and shipped around US$500 million worth of computer systems. The initial work of the site was the final assembly and test of systems for European markets. This involved configuring systems, testing and order consolidation. As computer systems became more reliable and easier for customers to configure for themselves, the traditional final assembly and test operation became redundant. So the plant in Scotland introduced the volume manufacture of the small "micro PDP 11" computer system for the European business market, with an additional £4 million investment, mainly in new production equipment. The product life of these new systems was short, with sales peaking over about three years. Product improvements were being introduced continuously. The market was extremely competitive and sales volumes were difficult to predict accurately. The plant at Ayr had to demonstrate that they could compete effectively on quality and cost with other Digital plants elsewhere in the world. The new product range involved changing from skilled technical configuration and test work to volume assembly, with no short-cycle times on individual operations. Existing staff were retrained, and there were no redundancies.

The management approach involved the following integrated package of related changes:

.../...

Case-study 5G (cont.)

Strategic focus

Management had a clear view of future products and of the design of organization that would be required to manufacture them competitively. This clarity of vision was shared with employees and helped to "sell" sweeping technical and organizational changes. From a strategic, or competitive, viewpoint the two outstanding advantages of the high-performance approach concerned the quality of the end products and the flexibility of the organization.

Supportive policies

A new skills-based payment system was designed, with employee participation. Career development was flexible, and management encouraged individuals to pursue their interests and extend their skills in areas where they could make a contribution to the business. Employment policies concerning rewards, training, development and career planning thus positively supported the strategic focus of the technical and organizational innovations.

Kit to fit

A capital-intensive, highly automated manufacturing process would have reduced unit costs only with high production volumes, which could not be guaranteed in an unpredictable market. An automated process would not be able to cope with rapid and fundamental changes in product design and technology. Manual assembly with stand-alone automated equipment was less expensive, less risky, more flexible – and was therefore a better fit for the business.

High-performance teams

Management introduced autonomous teams, each with around 12 members, with full "front-to-back" responsibility for product assembly and test, fault-finding and problem-solving, as well as some equipment maintenance. The group members used flexitime without clocks and effectively policed their own team discipline. Individuals were encouraged to develop a range of skills and to help others to develop their capabilities. The ten key characteristics of Digital's high performance teams are listed below:

- **self-managing, self-organizing, self-regulating;**

- **front-to-back responsibility for core process;**

- **production targets negotiated with management;**

- **multi-skilling and no individual job titles;**

- **shared skill, knowledge and experience;**

- **skills-based payment system;**

- **peer selection, peer review;**

- **open layout, open communications;**

- **support staff on the shop-floor, on the spot; and**

- **commitment to high standards of performance.**

.../...

Case-study 5G (cont.)

Management style

Management had to adopt a supportive style in their relationships with the high-performance teams. Some managers found it difficult to relinquish their traditional directive management style, and moved to other parts of the company. The teams initially had leaders whose job was to encourage team autonomy, and to develop their problem-solving and decision-making skills to the point where the team leader could withdraw. The need for patience, and the ability to stand back and let the groups reach their own decisions, became new management skills. Group decision-making was slow at first, but with experience, groups learned to diagnose and resolve problems rapidly and effectively. This approach released management time, which enabled management to devote more attention to suppliers, sales forecasting, operational issues and logistics, environmental scanning, new techniques, new ideas and to finding new business.

Supportive systems

Management introduced a computer-aided system for materials acquisition and production control. Support groups were located on the shop-floor, around the teams which they serviced.

Transition management

The process of managing the transition from a conventional manufacturing operation to one based on high-performance system concepts was critical to the success of the approach. One manager was primarily responsible for securing the new product line, and for persuading management colleagues that their novel approach to organizational design and management style was going to be effective. This involved a time-consuming combination of formal committees and informal conversations to encourage people to voice doubts and ideas. This process began early, which gave people time to absorb the new ideas, and the changes in behaviour and attitudes that were required.

Involvement

Communication with all those to be affected took place early and was sustained. A year before manufacturing began, a project team was set up including managers, engineers and assemblers. The high-performance team idea was explained through a programme which, again, relied on key personalities and regular meetings to spread the ideas. Management efforts in this respect were critical to the sustained enthusiasm and commitment of others. The ultimate test of this communications and involvement process was that staff eventually felt that they "owned" the concepts and techniques which they used, having contributed to the implementation.

Competence

Assessment of training needs was carried out early, and training was thorough – covering job skills, problem-solving techniques, and training in the concepts of high-performance work system design. The revised payment and performance review systems, which employees had helped to design, also assessed team members on their behaviours and attitudes as well as work output. Training on new equipment was initially carried out by suppliers, but skilled team members were then expected to train others, and particularly new team members.

.../...

Case-study 5G (cont.)

The company achieved a number of significant benefits from introducing this package of changes, including:

- high and sustained product quality;

- zero slippage on customer orders;

- outstanding energy;

- functional barriers and conflicts reduced;

- greater flexibility in responding to customers;

- reduced time to market for product innovations;

- improved communications with open process layout;

- better business understanding and priority setting;

- better identification with the product; and

- multi-functional career development possibilities.

The approach generated, in the words of one manager at Ayr, "massive personal growth and skills development" among employees, particularly with respect to:

- analysis and synthesis skills in problem diagnosis;

- interpersonal, self-presentation and communication skills;

- group problem-solving and decision-making skills;

- group self-management skills; and

- process design and planning skills.

The approach can thus become self-perpetuating through continuing employee skills growth. A review carried out in 1987 showed that these benefits had been sustained and, in 1989, the company introduced this approach to its semiconductor assembly activity at Ayr, with similar positive results.

Sources: Buchanan, 1989(a), pp. 255-273; Buchanan, 1989(b), pp. 78-100; and Buchanan and McCalman, 1989.

Features of autonomous working groups

Although autonomous working groups are implemented in different ways, according to local situations and needs, some key features can be identified. The following is not a comprehensive list nor one which one would expect to be always completely reflected in the concrete experience of autonomous working groups within companies (Lawler, 1986):

- allocation of whole sections of work to teams;
- self-managing teams with elected leaders;
- team responsibility for goal-setting and task allocation;
- team control over quality and absenteeism;
- production teams select new members;
- multi-skilling and skill sharing;
- skills-based payment systems with gain sharing;
- team reviews of skills and salaries;
- process layout supports teamwork and communications;
- flat management hierarchy – no foremen;
- some support functions carried out in teams;
- fewer support personnel in engineering, scheduling and quality;
- support staff become consultants/trainers;
- emphasis on training, career planning and personal development; and
- clearly expressed and communicated management philosophy.

Benefits and limitations of autonomous working groups

Autonomous working groups can bring important benefits to the enterprise and the people working in it. These benefits include:

- improvements in working methods and procedures as a result of team problem-solving;
- excellent results in attracting and retaining employees because of involvement and pay;
- cross-training leads to high levels of staffing flexibility and significant savings;
- quality is very high due to employee motivation to do high quality work;
- rate of output is good due to design and motivation;
- less staff support is needed because the teams incorporate the necessary skills;
- less supervision because teams are partially self-managing;
- low levels of grievance reporting because teams resolve issues themselves, quickly;
- good decision-making because of teams and widespread input of ideas;
- high level of continuous skill development;
- improved working conditions leading to higher level of job satisfaction reduces cost of absenteeism and labour turnover, thus increasing productivity;
- adaptability to new technology and change in general due to higher workforce flexibility;
- better response to customer needs;
- increased loyalty to the enterprise; and
- better environment.

However, their implementation can be accompanied by a number of problems, as follows:

- expectations for personal growth and development can reach beyond the ability of the organization to satisfy them;

- some people just prefer traditional ways of working;

- the team leader role can be ambiguous, with no role models to follow;

- group skills are needed for participative problem-solving;

- support activities often remain unaffected, and become jealous;

- "regression under pressure" to traditional management style;

- implementation timing – who should be involved and when?

- the pilot implementation which creates the "foreign body in the larger organization" problem;

- training costs are high because of cross-training and team training; and

- can meet with resistance from middle management.

Autonomous working groups represent an organizational culture change, not just a new work design technique. It can take two or more years to develop fully, and attitudes and behaviours have to change at all levels. The organization adopting this approach has to be prepared to take these considerations seriously and plan for them in implementation. They may thus not be appropriate for all organizations. In some cases, job rotation, enlargement and enrichment techniques may be more appropriate to address current issues and to solve current problems, and as a basis for more radical developments at a later date. The details of approach have to be designed to fit with local traditions and conditions. The timing of the implementation must also take into account existing work practices, the skills base and employee expectations. Change that is perceived as too radical and too rapid can create anxiety and frustration, and even anger. On the other hand, the same responses can be generated when management introduce change that is perceived as too limited and too slow. Analysis and judgement are required to resolve these detail and timing problems case-by-case and the involvement of all parties concerned.

References

D. Boddy and D. Buchanan: *The technical change audit: Action for results*, Module No. 5 (Sheffield, Manpower Services Commission, 1987).

D.A. Buchanan: *The development of job design theories and techniques* (Aldershot, Saxon House, 1979).

— [1989a]: "High performance: New boundaries of acceptability in worker control", in S.L. Sauter, J.J. Hurrell and C.L. Cooper (eds.): *Job control and worker health* (Chichester, John Wiley, 1989), pp. 255-273.

— [1989b]: "Principles and practice in work design", in K. Sisson (ed.): *Personnel management in Britain* (Oxford, Basil Blackwell, 1989), pp. 78-100.

— and J. McCalman: *High performance work systems: The Digital experience* (London, Routledge, 1989).

P.F. Drucker: "The emerging theory of manufacturing", in *Harvard Business Review*, Vol. 68, No. 3, May-June 1990, pp. 94-102.

B. Dumaine: "Who needs a boss?", in *Fortune*, No. 10, 7 May 1990, pp. 40-47.

F.E. Emery: "Some hypotheses about the ways in which tasks may be more effectively put together to make jobs", in P. Hill: *Towards a new philosophy of management* (Aldershot, Gower, 1971), pp. 208-210.

European Trade Union Institute: *Redesigning jobs: Western European experiences* (Brussels, May 1991).

C. Gallagher and W. Knight: *Group technology* (London, Butterworth, 1973).

D. Grayson: *Self regulating work groups: An aspect of organizational change*, Occasional Paper No. 46 (London, Work Research Unit, July 1990).

P.G. Herbst: *Autonomous group functioning* (London, Tavistock Publications, 1962).

L. Hirschhorn: *Beyond mechanization* (Cambridge, Mass., MIT Press, 1984).

J. Hoerr: "The payoff from teamwork", in *Business Week*, 10 July 1989, pp. 56-62.

—, M.A. Pollock and D.E. Whiteside: "Management discovers the human side of automation", in *Business Week*, 29 September 1986, pp. 60-65.

Industrial Relations Services: *European Industrial Relations Review* (London), Vol. 210, July 1991.

—: *European Industrial Relations Review* (London), Vol. 217, February 1992.

ILO: *New forms of work organization*, 2 volumes (Geneva, 1979).

—. Management Development Branch: *Field experiments in new forms of work organization* (Geneva, January 1981).

G. Kanawaty: *Introduction to work study* (Geneva, ILO, fourth revised edition, 1992).

M. Kirosingh: "Changed work practices", in *Employment Gazette*, August 1989, pp. 422-429.

K. Kogi, W.O. Phoon and J.E. Thurman: *Low-cost ways of improving working conditions: 100 examples from Asia* (Geneva, ILO, 1988).

A.R. Kroll: "New working practices for blister packaging lines", in *Manufacturing Chemist*, November 1989, pp. 33, 35 and 38.

E.E. Lawler: *High involvement management: Participative strategies for improving organizational performance* (San Francisco, Jossey-Bass, 1986).

S.A. Mohrman, G.E. Ledford and E.E. Lawler: "Quality of worklife and employee involvement", in C.L. Cooper and I. Robertson (eds.): *International Review of Industrial and Organizational Psychology* (New York, John Wiley, 1986), pp. 189-216.

N. Oliver: "The dynamics of just-in-time", in *New Technology, Work and Employment*, Vol. 6, No. 1, 1991, pp. 19-27.

— and B. Wilkinson: *The Japanization of British industry* (Oxford, Basil Blackwell, 1988).

J. Parnaby: "A systems approach to the implementation of JIT methodologies in Lucas Industries", in *International Journal of Production Research*, Vol. 26, No. 3, 1988, pp. 483-492.

B. Perry: *Enfield: A high performance system, Digital Equipment Corporation* (Bedford, United Kingdom, Educational Services Development and Publishing, 1984).

T. Peters: *Thriving on chaos: Handbook for a management revolution* (London, Macmillan, 1987).

M. Piore and C. Sabel: *The second industrial divide: Possibilities for prosperity* (New York, Basic Books, 1984).

R.B. Reich: *The next American frontier* (New York, Times Books, 1983).

A.K. Rice: "Productivity and social organization in an Indian weaving shed", in *Human Relations*, Vol. 6, No. 4, 1953, pp. 297-329.

—: *Productivity and social organization* (London, Tavistock Publications, 1958).

R.J. Schonberger: *Japanese manufacturing techniques* (New York, Free Press, 1982).

—: *World class manufacturing: The lessons of simplicity applied* (New York, Free Press, 1982).

B. Stevens: "Integrated working at Blue Circle", in *Industrial Participation,* No. 594, Summer 1987, pp. 10-12.

Swedish Work Environment Fund: *New job content produces good results* (Stockholm, 1989).

E.L. Trist and K.W. Bamforth: "Some social and psychological consequences of the longwall method of coal-getting", in *Human Relations*, Vol. 4, No. 1, 1951, pp. 3-38.

—, G.W. Higgin, H. Murray and A.B. Pollock: *Organizational choice: Capabilities of groups at the coal face under changing technologies* (London, Tavistock Publications, 1963).

R.E. Walton: "From control to commitment in the workplace", in *Harvard Business Review*, March-April 1985, pp. 77-84.

— and G.E. Susman: "People policies for the new machines", in *Harvard Business Review*, No. 2, March-April 1987, pp. 98-106.

P. Wickens: *The road to Nissan: Flexibility, quality, teamwork* (Basingstoke, Macmillan Press, 1987).

S. Zuboff: *In the age of the smart machine: The future of work and power* (Oxford, Heinemann Professional Publishing, 1988).

WELL-PLANNED BUILDINGS AND PREMISES

6

Juan Carlos Hiba

This publication has up to now been discussing how to design work to fit your business. Now it is also important to pay attention to where you will carry out the production, and how the premises and facilities should be arranged to make the production effective and efficient.

Why is it so important to give special consideration to the building, to premises and to its characteristics?

Buildings are more or less permanent constructions. They are normally expected to last many years, 50 years and more; they should provide shelter for people working inside; they should provide an appropriate space and volume to hold the technology; they should be flexible enough for adaptation to changing trends and new products; they should comply with government regulations; and, above all, they cost a lot of money. In summary, to build new premises requires a great deal of resources and time, and you will be trying to satisfy these many factors to get the best building for the money.

Siting the industrial enterprise

Availability of land is obviously important. There will probably be government requirements involving town planning, industrial zoning and environmental regulations. Associated with these are the provision of services, such as power and water, as well as drainage and other emissions which may be noxious to local people.

Then there are questions of logistics, availability of supplies or subcontractors, nearness to markets, the transport facilities available and access to them. Associated with these is the availability of people to work in the plant, the education facilities in the neighbourhood and the region, the accessibility of the plant to where people live and the public transport available.

Different types of factory premises

The premises designed or chosen depend not only on the size of the company, but also on the technology being used in production. Decisions also have to be taken concerning the human requirements for an effective plant. These have to be considered right from the start, because if building is begun and equipment ordered in advance of studying the ergonomic and organizational requirements, then it could well be that obstructions to good performance will be built into the premises, and be very expensive to change later.

To do this ergonomic planning, the designers have to work from the proposed manufacturing sequence and methods, and be clear about all the functions which each person is expected to fulfil. A task analysis, as described in Chapter 2, is then assessed in conjunction with the proposed equipment, to see what ergonomic requirements may have to be specified to be sure that each person can do the required activities effectively. These decisions will, in some cases, affect the layout as well as the equipment specification, so they are done at an early stage, and run in parallel to the initial layout decisions.

Decisions on equipment will require it to be evaluated for its useability and safety. Noise and vibration are two important aspects: the build-up of noise when several machines are together can lead to excessive noise levels in a plant. Where the process itself is noisy, e.g. cold heading of bolts, either the machines should be enclosed or put in an enclosed space, with double-glazed windows being provided so that operators can see the process but only spend a minimum amount of time exposed to high noise levels where they will need to wear ear muffs.

Basic recommendations on other aspects of the physical environment are given in Chapter 3, where values for lighting, temperature, etc., are noted.

The arrangement of manufacturing equipment is discussed in Chapter 4. When considering this in relation to the building, arrangements for material movement and maintenance are important factors, including access, space and job aids such as cranes or fork-lift trucks.

Table 6.1 shows a scheme of a generalized design process including the social, human and environmental (SHE) factors that should be taken into account right from the start when new buildings and industrial plants are being planned and designed.

Individual offices or open-plan offices?

If you are involved in designing new premises or if you are selecting ready-to-use offices for your enterprise, one strategic decision to be made concerns the facilities you want for your staff.

It is usual to provide individual offices for your senior staff, but for other office workers you have two main choices: either individual offices or an open-plan office. Both have advantages and disadvantages. Several factors should be taken into consideration when making a decision on what type to adopt. Table 6.2 may help you in making an appropriate decision.

Although these comparisons suggest that, in principle, the open-plan office is better, it is important to consider that people differ in their need for a personal space and in their response to being always on view. Hence, it is usual to provide some partitioning, helping to indicate functional divisions between different office activities as well as forming recognizable work groups which can mix and discuss without disturbing others.

Table 6.1   An example of a design process in which the social, human and environmental factors are considered

PHASE	0	1	2		3	4	5
Production process	Definition	Function analysis	Layout and selection of processes and machinery	Design of hardware and purchase specifications	Analysis	Installation	Operation
	of				of	and	Evaluation and testing
	system				bids	construction	
S H E factors	objectives	Definition of operational requirements for S H E	Design assistance		and the purchasing		
	and				processes		In use
Building	general	Analysis	Design				
	criteria		Function design	Physical design	Detailed design		

| Examples of S H E factor activities | Definition of activities
– Selection of site
– Studies of other plants and buildings | To give requirements on:
– Noise
– Light
– Thermal environment
– Ventilation
– Waste (gases, liquids, solids)
– Social norms
– Social space
– Communication flows
– Job principles | – Selection of machinery (safety, vibration, layout, instruments)
– Selection of building techniques (acoustics, lighting, ventilation plan, physical communication)
– Development of control routines, information programmes, organization | | | Advice to management and personnel departments on selection, training, safety programmes | |

Source: Ivergård, 1973.

Table 6.2 Traditional partitioned offices versus open-plan office

Factor	Traditional office partition	Open-plan office
Communication and collaboration	Easy between colleagues working in the same office; more difficult for staff working in different offices	Easy access to all staff
Disturbances and noise level (phone, dictating, visitors)	Mostly over 60 db(A); higher interference from telephone calls and visitors in shared small offices	Mostly below 60 db(A) due to more favourable acoustics; background noise can mask conversation
Illumination	Daylight prevailing; big differences of illumination throughout office; possible glare from windows	Daylight available only for desks close to windows; artificial but more homogenous distribution of lighting; possible glare from bulbs
Climate; ventilation	Ventilation by opening windows if air conditioning not available; easy adjustment to comfort of the individual if few people in the office; smokers affect fewer staff	Air conditioning necessary; individual adjustment difficult; disagreement concerning temperature may arise; smokers affect more staff
Flexibility in arranging workplaces	Considerably reduced due to space limitation	Important flexibility in office layout and design of individual workplaces; layout easily adaptable to changes in the organizational structure
Efficient use of space	Less efficient; losses due to space for individual entrance and corridors and walking space; furniture and walls creating corners not being used	More efficient; less individual corridors; flexible use of common available space
Psychosocial effects	Furnishing and design of office to suit individual taste (choice); office size and equipment accentuate rank/status differences; conflicts among colleagues sharing one office may arise; supports the feeling "my office is my castle"	Consistent overall design concept but less "personalized"; rank/status differences less visible; individual feeling of right control by other colleagues and supervisors; teamwork easier to arrange
Visual contact with external world	Ensured for almost all staff	Ensured only for staff with seating near to windows
Information exchange; work flow	Encourages "batch work" operations; more difficult information exchange between colleagues in different offices; tends to contribute to longer throughput time	Encourages continuous work flow; easier information exchange throughout office; tends to contribute to shorter throughput time

Source: Nixdorf Computer AG, 1989.

VDU workstations

Workstations equipped with visual display units like word processors and other information technology equipment are very common. Several ergonomic factors should be taken into consideration if you want to have very productive workplaces. Apart from the appropriate design of the workstation discussed in Chapter 3, the correct positioning of the VDU unit with respect to windows and other light sources is very relevant. If the VDU unit is positioned so that the user is facing a window, direct glare will affect the operator. Turning the VDU unit 180° so that the worker has his or her back to the window does not solve the problem either. Reflected glare on the screen will occur.

The following illustrations show three choices for positioning workstations equipped with visual display units, depending on the number of walls with windows in the building. Figure 6.1 shows the most favourable positioning of workstations with VDUs when only one wall contains windows.

Figure 6.1   Positioning of standard (non-VDU) workstations closer to the windows. VDU workstations are distributed in rows, but far from the windows. In all cases, workstations neither face nor back on to the windows

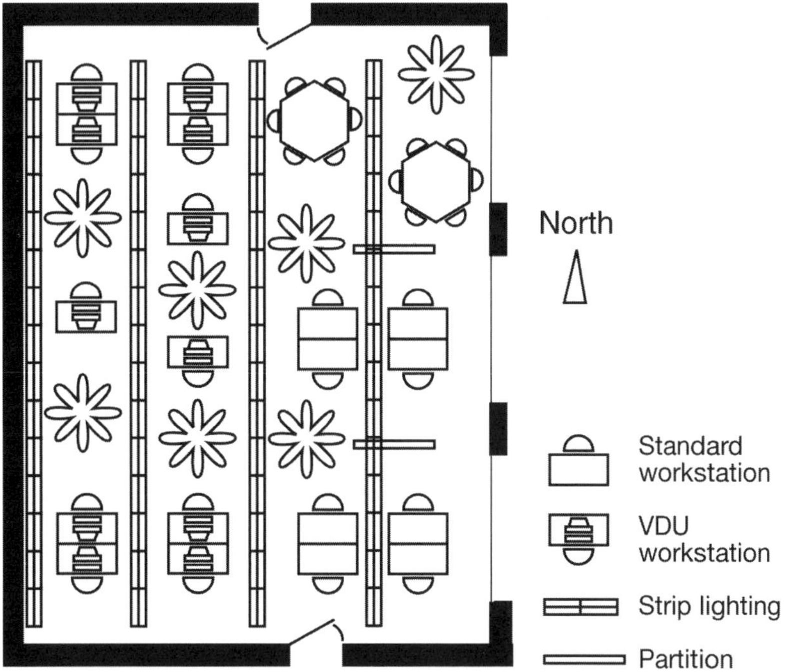

Source: Benz et al., 1983.

Figure 6.2   Distribution of VDU workstations and standard workstations when one set of windows faces south and two adjoining walls contain windows. Curtains may also be needed on the east façade

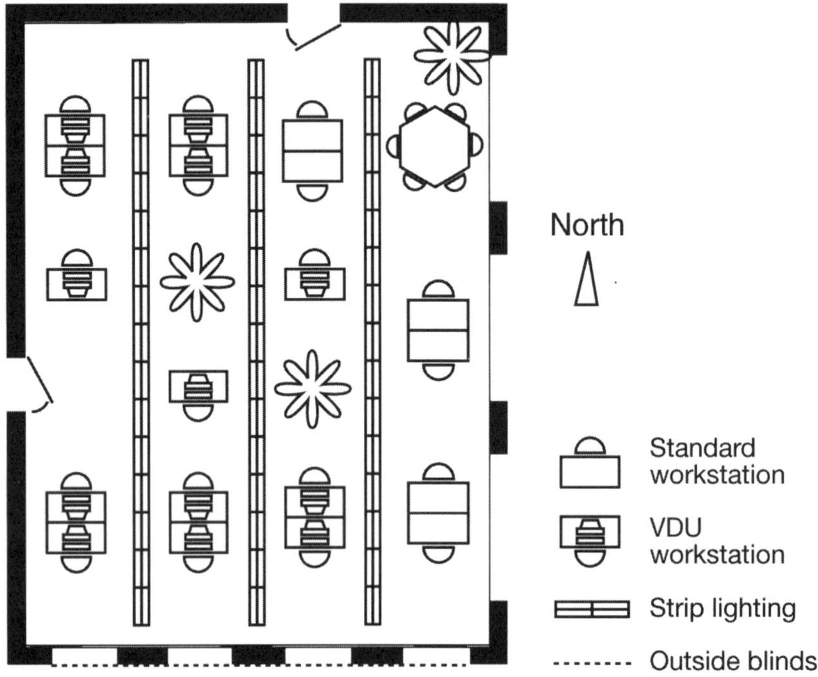

Source: Benz et al., 1983.

Figure 6.3   Recommended position of VDU and standard workstations when there is a profusion of windows in three façades

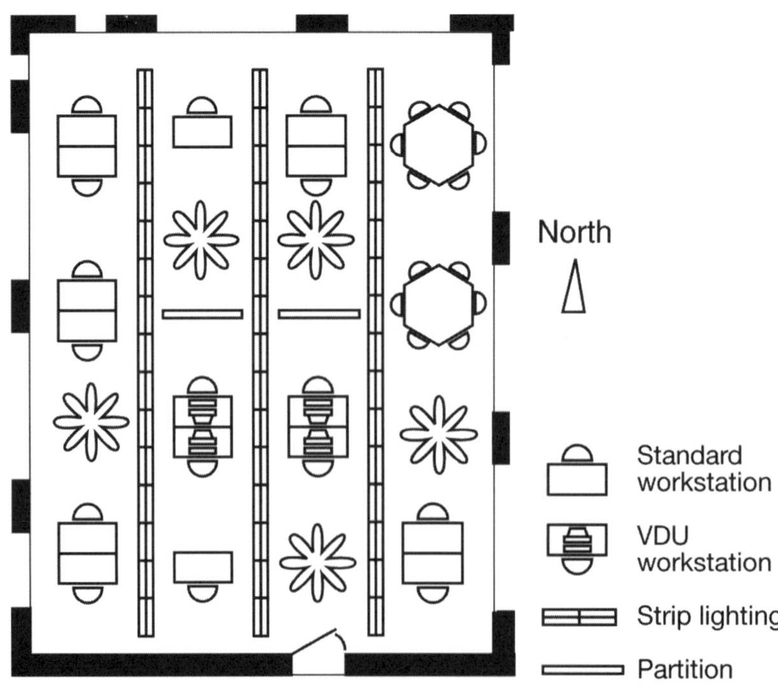

Source: Benz et al., 1983.

When the office has windows facing south, blinds or curtains should be provided to control the sun's radiation and direct glare. Depending on the window height, low partitions may also contribute to avoiding glare from bright windows. Figure 6.2 illustrates this case.

There are cases where windows are installed massively and in three facades. While standard workstations and working tables should be distributed near windows, workstations equipped with word processors should be located preferably in the centre of the room.

Note that all these recommendations apply to the Northern hemisphere and, in general, to the temperate zones. They can be revised for the Southern hemisphere, but in tropical areas shading of some kind, shutters or blinds, will be needed to deal with sunlight at different periods of the year, as well as with the more intense light experienced in those regions. In many parts of the world, traditional building designs have evolved to cope with the local conditions, and this experience should not be ignored.

Artificial lighting and the VDU workstation

Light reflections from ceiling lights can also create visual problems for operators. Such reflected glare is a particular problem when the ceiling lights are very low or when the ceiling lights cross the line of sight of the operator. This can be controlled to some extent by locating the workstations between two rows of lighting units, while lights should be shaded so that the light source is not visible from eye level. Provision of partitions can also prevent reflections from light fittings as well as from windows.

Assessing the environmental factors
in VDU workstations

Lighting, climate, acoustics and workstation configuration are the four main issues that can affect working conditions and therefore productivity.

The following checklist (box 6.1) can help you assess the situation of your VDU workstation. For an answer to more than half the questions, you need no measuring devices. For the rest, you may need some help. Do not forget to involve your employees. They can help you answer many questions.

Once you have applied the checklist, you can proceed to analyse it with the cooperation of the people who work in the environment. You will usually find a series of immediate changes that could be implemented for improving working conditions.

Box 6.1   Checklist for assessing environmental factors in VDU workstations

Lighting

* **Is the horizontal illuminance around 500 lx?**
* **Are the following reflectances observed?**

Ceiling	**around 70%**
Walls	**around 50%**
Partitions	**around 30 to 50%**
Furniture	**around 20 to 50%**
Curtains	**around 50 to 70%**

* **Is the user's field of sight free of sources of disturbance (lamps, windows)?**
* **Is the screen free of disturbing glare and reflections?**
* **Is the lighting equipment serviced regularly?**

Climate

* **Is the room temperature around 21°C (max. 26°C)?**
* **Is the relative humidity between 40 and 65%?**
* **Is the air velocity no higher than 0.15m/s?**
* **Is intense solar radiation prevented by external blinds (especially for windows facing the sun)?**
* **Have measures been taken to ensure that employees do not sit in the flow of air from the cooling system of the equipment?**

Acoustics

* **For predominantly mental activities, is the noise level no higher than 55 dB(A)?**
* **For normal office activities, is the noise level no higher than 70 dB(A)?**

Configuration of workstations in the room

* **Is the user's line of sight parallel to the windows?**
* **Are shielding measures provided for when windows in the line of sight cannot be avoided?**
* **Are shielding measures provided for when windows behind the user cannot be avoided?**

Source: Benz et al., 1983.

VDU workstations and techno-stress

Under the pressure of increasing "demands" (justified by the supposedly facilitating function of new technologies), VDU users often find it difficult to "control" the content and quality of their work. This gap between "demand" and "control" is the cause of their stress.

Closing the gap is for the future. However, all information technologies have an in-built flexibility which, contrary to widespread deterministic views, allows for adaptability to different circumstances and needs. This means that there is always room for manoeuvre and that different solutions are available, both technically and organizationally.

In this context it is certainly possible to activate dialogue at an early stage between software designers, management and workers in order to test a number of different solutions for introducing VDUs in the workplace, to assess their impact and to introduce them in a participatory way.

Participatory design thus becomes an essential prerequisite to the introduction of VDUs at the workplace. It is also important that, in order to avoid techno-stress generated by poor social contacts and isolation, participation is developed as a core element of VDU use. An effective way of enhancing the role of participation in relation to work on VDUs is by introducing teamwork. Users can thus share work, experience and information with other users in their immediate work environment. Teamwork in offices would facilitate the integration of VDUs, improve relationships with colleagues and management, and, eventually, the quality of overall performance.

Job design is another essential element in this respect. An efficient design should provide for an appropriate variety of skills, capacities and activities; ensure that the tasks performed are identifiable as whole units of work rather than fragments; that these tasks make a significant contribution to the total function of the system which can be understood by the user; that they provide an appropriate degree of autonomy, sufficient feedback on task performance and opportunities for the development of staff skills. Within this context, the potential for job enlargement and job enrichment should be given priority attention.

VDU users would thus progressively move from a traditional, limited range of tasks, such as typing, document editing and data entry, into a more varied and complex combination of functions, including electronic classification, filing and search of documents, self-training and training of colleagues, up to the direct support of managerial activities.

The development of VDU activities through time also deserves careful consideration. Computer technology allows for full determination and control of the pace at which users have to work. They may thus by subject to undue time pressure, long working periods without intervals or prolonged waiting periods. This may, in turn, generate fatigue accumulation and stress as well as sensations of monotony, boredom and dissatisfaction.

Working time design is therefore an essential component for an efficient organization of the VDU user's work. The duration and distribution of time spent operating the system is also crucial in reducing possible negative effects of VDU work on the user's health. The work should be split preferably into several short periods rather than a single long period on the screen. A mixture of VDU and non-VDU work is recommended; short breaks are also necessary. The optimal length of pauses will be related to the nature of the task. The effectiveness of the pause will also be a function of when it is taken. In general, rest pauses should be taken prior to the onset of a noticeable fatigue. Short, frequent pauses appear to be more efficacious than fewer, long pauses. Worker-controlled and informal rest pauses may be preferable to rigid, supervisor or process-controlled rest pauses.

System design is the final and most important factor to be considered. It is also the most difficult to control by the users, since the producers often predetermine the entire setting and the operational functioning of the system. However, every effort should be made to explore, particularly when purchasing new software, its real capacity of response to the user needs with special attention to the risks of techno-stress for VDU users. The system may be too demanding or too slow, it may be too rigid or too loose, it may be a channel for excessive control on the staff, or require from the user a continuous high level of attention which may lead to unnecessary strain or fatigue. It is indeed necessary to have a system which has a large built-in flexibility, adaptable to varying circumstances and needs, easy to control and capable of facilitating ongoing planning and full communication. This is altogether an open system, focused on the user's concerns, which is able to develop a balanced interaction between individual, organizational and technical requirements in such a way as to positively affect the performance, health and well-being of the worker concerned.

Floors

We are inclined to underestimate the importance of the floor for productive, smooth and safe work. However, inappropriate floor surfaces or poorly maintained floors can be a major source of accidents, work interruptions and product damage.

The most important qualities of the floor surface are:

–   **Strength.** A floor should be sufficiently strong to resist crushing due to loads from heavy machines or from traffic or handling of materials. Care must be taken to ensure that the wheels on mobile materials-handling equipment are of adequate width and diameter;

–   **Resistance to wear and abrasion.** The floor must have sufficient resistance to abrasion to withstand normal use over a period of several years without deterioration and without excessive signs of local wear. Non-dust-forming properties are important and can be critical for certain industries (electronics, food, etc.);

–   **Resistance to chemicals.** It is important that the floor should be resistant to chemicals wherever there is a risk of oils, solvents, acids or other chemicals being spilt. This applies particularly in the chemical and petrochemical industries;

–   **Comfort and safety.** The floor should have low thermal conductivity and absorb noise and vibration as well, since these phenomena have a direct effect on the occurrence of fatigue. A machine operator standing the whole day on a concrete surface gets much more tired than a worker on a wooden one. Furthermore, the floor should always help to avoid slipping and be easy to clean.

When considering the cost factor, we should not think exclusively in terms of initial installation but also take into consideration long-term durability, easy maintenance and cleaning.

Table 6.3 can help in the selection of appropriate material for the workshop floor.

Floors which are frequently washed down with water should have a slight, even grade of 1 to 2 per cent towards a drain to ensure that the water flows away from the traffic area.

Table 6.3 Floor characteristics

Properties	Clay	Concrete*	Wood blocks*	Poured asphalt	Plastic (vinyl)	Ceramic
Strength						
Compression resistance	—	++	+	+-	+-	++
Impact resistance	+	+-	++	+	+	-
Resistance to wear and abrasion						
Dust formation	Yes	Yes	Yes	No	No	No
Resistance to chemicals						
Water resistance	—	+	-	++	+	++
Acid resistance	+	-	+	-	+	++
Alkali resistance	+	+	+	+	+	++
Oil resistance	+	-	+	—	+	++
Solvent resistance	+	+	+	-	+	++
Comfort and safety						
Thermal insulation	++	-	++	+-	+-	-
Ease of cleaning	—	+-	-	+	++	++
Dielectric properties	—	-	+[1]	+	+	+
Friction sparking	No	Yes	No	No	No	Yes

++ = very good
+ = good
+- = medium
- = bad
— = very bad
[1] if dry

____

* Note that sealing treatments are available for these surfaces.

Source: Thurman et al., 1988.

Floors, falls and accidents

The material used for floors has a relevant influence on accidents. Independently of other physical characteristics, their property for impact attenuation is important in human accidents involving falls.

Experimental results (Maki et al., 1990) suggest that the nature of the floor covering may play a substantial role in impact attenuation for falls on the hip. In many countries, hip fracture is the most common injury leading to hospital admission.

Table 6.4 lists the 13 floor coverings tested with simulated "falls" on a test rig.

Table 6.4 Testing of different floor coverings

Floor covering	Peak deceleration (g)		Impact attenuation
	Mean	Standard deviation	[qualitative][1]
Thick pile carpet + pad	15.32	0.64	very good
In/out carpet + pad	15.81	1.12	
Pile carpet + pad	15.96	0.21	
Loop carpet + pad	16.60	0.71	
Thick pile carpet	17.74	0.57	
In/out carpet	17.79	0.57	good
Loop carpet	18.00	0.40	
Wood tile	18.99	1.25	
Terrazzo	19.33	2.31	
Vinyl	19.98	2.13	
Vinyl composition	20.11	1.43	
Linoleum	20.46	1.41	regular

[1] Note: Impact attenuation characterization is qualitative and refers to comparison between extremes.

Source: Makie and Fernie, 1990.

Direction-finding systems for identification of workplaces

Whether the building you are planning or designing, or the premises where you are now operating, are complex or simple, it is always safer and useful to have a direction-finding system installed.

A signalling system will help workers to save their productive time. Occasional visitors will also arrive at their target department, office, etc., without delays and without getting lost in a hostile environment.

Direction-finding systems depend upon the design of the building concerned and of the major and minor destinations that people want to find. The following general method for developing a direction-finding system is proposed (Corlett et al., 1972):

1. Identify major destinations and common routes.

2. Determine the most direct route between each building entrance and major destinations. Consider also minor or less frequent destinations.

3. Identify the route choice points.

4. Ascertain the most common or familiar search terms for various destinations by interviewing building employees. In the case of public buildings, carry out an enquiry with the building users.

5. Specify the direction-finding information required at each identified route choice.

6. Group destinations showing common routes.

7. Prepare a building directory which relates minor destinations to the routes used for major destinations.

Designing the finding system requires decisions concerning the identification of general and specific building areas to be reached; restricted areas; route choices; places for positioning frames or labels; board size; size and shape of symbols and arrowheads; size, colour and contrast between text and background. The following design guidelines will help you.

For the design of an efficient signboard system, consider the following three distinct aspects of the system which will aid its use:

– Develop "generic search words" to enable direction finders to identify their objectives within the system.

– Provide "instructive symbols", such as arrows or signs, indicating direction; these favour the selection of the appropriate direction-finding response.

– For maximum clarity use good contrast between letters and background; use simple lower-case letters, not a decorative style; use simple solid symbols and be consistent throughout the sign system.

Box 6.2 Design guidelines for direction-finding systems

1. Design direction finders to first indicate the general area of their objective.

2. Signs should provide more precise or detailed information as they approach the destination.

3. Signs should follow the most direct route between their place of origin and their destination.

4. Install signboards before reaching the point of route choice.

5. Place sign boards perpendicular to the path of movement.

6. Install sign boards at the entrance in public waiting spaces, in lift lobbies, at lift exits as appropriate.

Source: Corlett et al., 1972.

Adapting worksites to workers with physical disabilities

Well-planned buildings and premises mean that they are adapted to the working people who will use them. Around 5 per cent of the population suffer some type of disability. Since people are different, therefore different workplaces should also be adaptable to the special aptitudes or circumstances of individuals, to suit also people with a disability or impairment.

Adaptation of the working environment to workers with some physical limitations does not necessarily mean huge investments or special, sophisticated equipment. There are always low-cost solutions at hand to enable a disabled worker to make an efficient contribution to production.

The following simple checklist (figure 6.4) can help in identifying accommodation required for adapting a workplace to a worker with disabilities.

Once the work requirements have been identified, the adaptation of the workplace requires the following steps:

Step 1. Assess the main job characteristics.
Examine the technical, physical and psychological requirements of the job's duties. Involve workers in this examination.

Step 2. Assess obstacles representing limits to employment. Determine functional disabilities by specifying how they represent obstacles in employment. Use checklist in figure 6.4.

Step 3. Determine what practical improvements can be implemented to adapt the workstation or the premises to disabled or impaired workers. Develop, as much as possible, simple, low-cost solutions. Investigate the technical and economic factors required for implementing the desired measures.

Step 4. Prepare an action plan.
Describe which measures can realistically be achieved, assign responsibility, allocate funds and establish a deadline.

Step 5. Implement the action plan.
Encourage everyone to start the adaptation process. Involve interested workers and co-workers. Follow-up the implementation process. Visit the workstation during the transformation period.

Step 6. Evaluate results.
Follow-up and determine the ongoing utility. Evaluate the adaptive measures after a period of time. Involve the workers in evaluating the results. Explore new opportunities for further adaptations.

Figure 6.4   Occupational requirements checklist for workers with disabilities

SWEDISH BOARD OF OCCUPATIONAL SAFETY & HEALTH
Occupational Medicine Department

DESCRIPTION OF OCCUPATIONAL REQUIREMENTS
Date

Company	Department	Analyst
Position	Work hours	
Salary conditions	Working position ☐ Sitting ☐ Standing/walking ☐ Alternating	
Job description		

	Assessment — Profile	Can be made easier	Comments
	1 2 3		
1. Sight			
2. Hearing			
3. Back			
4. Legs & feet			
5. Arms & hands			
6. Physical stamina			
7. Skin			
8. Noise			
9. Vibrations			
10. Climate			
11. Air pollution			
12. Supervision			
13. Teamwork			
14. Training time			
15. Accident risk			
16. Other critical factors			
Notes			

Source: Elmfeldt et al., 1983.

Worksite accessibility for workers with disabilities

Special measures should be observed in the adaptation of worksites for persons with various disabilities. The problems that are most common for the various disability groups and what measures can be taken are described below (Elmfeldt et al., 1983).

Mobility impairment

People with difficulty in movement often use sticks or crutches to steady themselves. For these individuals, it is difficult to walk long distances, climb steps, or open heavy doors. They require level, non-skid pavements and floors. Slack carpeting and isolated objects can create risks of stumbling and should be removed.

People with reduced strength in the arms and hands experience difficulty in opening doors, gripping and manipulating door knobs and fittings, etc. Door openers on heavy doors as well as easily-gripped handles and fittings can, therefore, be required.

For the wheelchair user, general mobility is the greatest problem. Worksites should, therefore, be sought which are generally close to the entrance. For the mobility-impaired individual, it is crucial that the doors and other passageways are sufficiently wide and that the locality is generally adequately spacious to permit manoeuvring with the technical aid such as wheelchair, crutch or stick.

Visual impairment

For those persons who have impaired vision, it is difficult to move about within various environments, as they cannot, or only to a limited degree, use their sight as assistance in orientation. Blind people orient themselves primarily with their tactile and auditory senses. Variance markings in the surface structure, for example markings on walkways, can be employed to assist blind people with mobility impairment. On exterior walkways, kerbstones and handrails are helpful for the blind person. Automatic revolving doors are difficult to negotiate.

Some visually impaired people orient themselves in part with their residual sight. In addition to the use of contrasts in surfaces and textures, colour contrasts and suitable lighting are also important in facilitating orientation and movement. For example, doors, fixture mountings, stairways, ramps, walkways, etc., should be in colours which contrast to the surroundings. Obstacles, in the form of pillars, protruding corners, isolated steps, protruding hat racks, etc. should be denoted in a special manner utilizing lighting, colour or texture contrasts.

Glass walls and automatic sliding doors should have a contrasting horizontal colour field, frame or similar marking directly at eye level.

Blind people may read by touch. Relief text is, therefore, required. Important information should not only be provided in written and Braille form, but also provided verbally, for example, in a factory setting via an intercom system.

Hearing impairment

People with reduced hearing have difficulty in distinguishing and comprehending sounds and speech in noisy environments. Noise-free environments and environments with good acoustics should, therefore, be sought. Important information should not only be provided in verbal form, i.e. via intercom systems, but also in written form. Many people with reduced hearing read lips. Lip reading is facilitated by good lighting. In order to assist the hearing capacity of those who use hearing aids, an induction loop installation might be needed at the worksite.

Allergy affliction

People with allergies might react to such substances as spores, dust, pollen or animal fur. Additionally, many are sensitive to direct contact with various substances and materials, such as nickel, chrome or rubber. Dust-collecting materials and textiles should, therefore, be avoided. Flooring and walling materials should be easy to clean. In addition, contact with allergy-causing substances should be avoided, for example on taps, door handles, hand rails and similar contact surfaces. Particle board, building board and furniture containing formaldehyde are examples of unsuitable building materials.

Developmental disability

Generally speaking, simple and logical facility layout, clear and single-step directions, as well as simple and logically designed control units, facilitate the functioning and working of developmentally disabled people.

Routes to and from the workplace

Pedestrian walkways to the workplace should be distinguishable from vehicle traffic ways, but they should, nevertheless, be connected to car garages, parking places and the stopping sites of public transport.

Irregular intervals in the surface levels should be avoided and inclinations should be no more than 1:50. If the walkway is shorter than six metres, the gradient should be no greater than 1:12.

In order for a wheelchair to be turned 180°, a squared space of 1.5m x 1.5m is needed. By varying the surface roughness of the walkway pavement, orientation for the visually impaired is facilitated, but it may make wheelchair use more difficult.

Unexpected changes in the walkway's layout should be marked, as should sudden surface level changes, ramps and stairs. The warning notices can be presented with signs as well as with different colour, form and texture contrasts. Different materials produce varying degrees of echo and can be useful to those persons who orient themselves with the assistance of an orientation cane.

Kerbstones along the walkway can facilitate orientation. The walkway should have uniform and non-glare lighting. It is particularly important that the lighting be suitable in places where the walkway changes character.

Handrailings should accompany moderate or steep slopes as well as longer passageways (figure 6.5).

Exterior steps should preferably be in a direct course. Isolated steps should be avoided or replaced by a short ramp. Long flights of steps should be divided by an intermediate landing. The step height and depth should be consistent throughout the entire course.

Exterior ramps may be built in order to negotiate minor changes in the level of the walkway and to supplement steps (figure 6.6).

Figure 6.5　Handrail dimensions for workers with disabilities

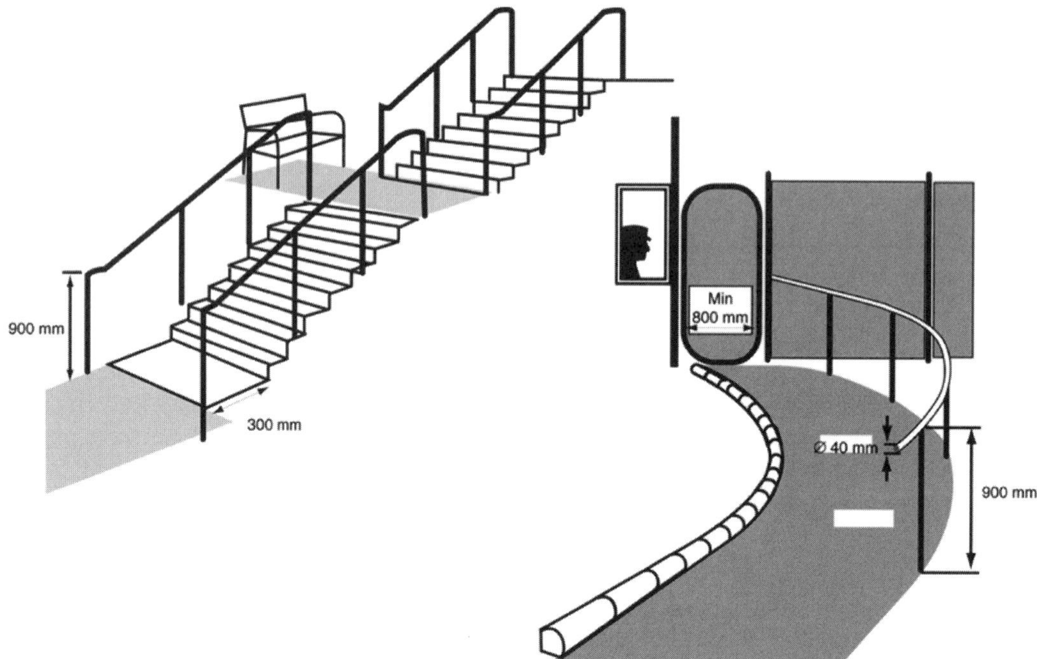

Source: Elmfeldt et al., 1983.

Figure 6.6　Ramp requirements for workers with disabilities

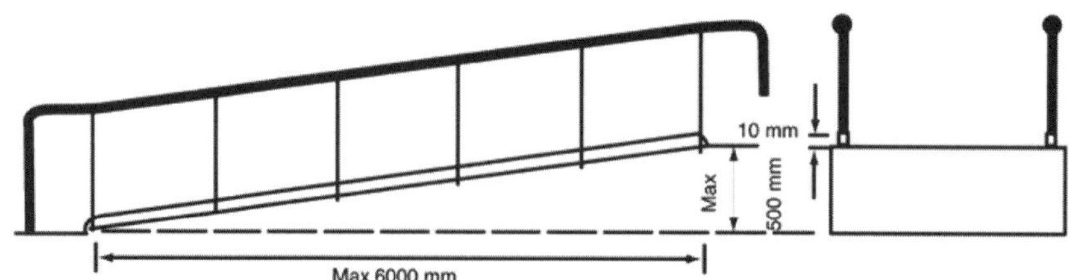

Source: Elmfeldt et al., 1983.

Entrances

At least one entrance shall be usable by persons with limited movement or orientation ability. The entrance must be easy to locate, close to the parking place and clearly marked (figure 6.7).

Figure 6.7 Entrance requirements for workers with disabilities

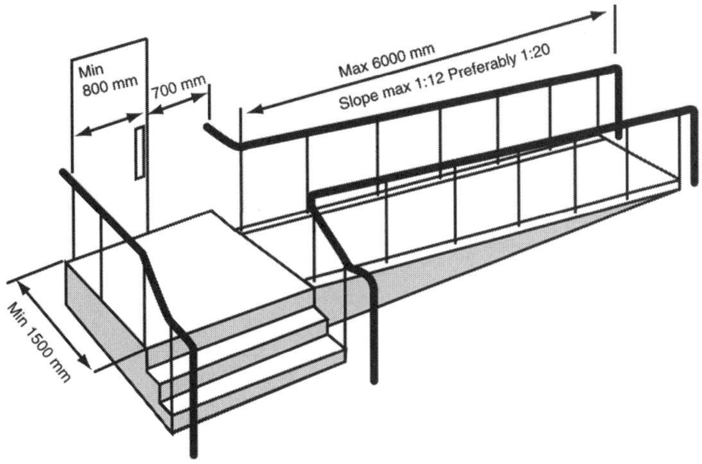

Source: Elmfeldt et al., 1983.

Parking places

The parking places for the disabled should be located as close to the entrance as possible. The road between the parking place and the entrance should fulfil the requirements designated for pedestrian walkways (figure 6.8).

Figure 6.8 Parking spaces for workers with disabilities

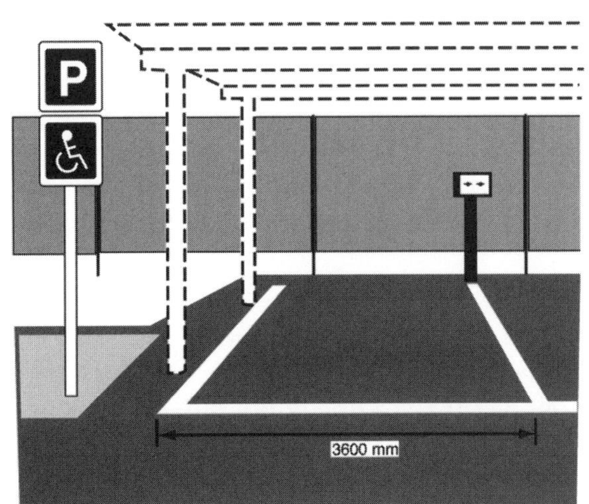

Source: Elmfeldt et al., 1983.

Transportation routes within the workplace

Lifts (elevators)

As regards new constructions, lifts are generally required for displacement between the various floors (figure 6.9). In adapting an existing workplace for an individual disabled worker in which the installation of a lift would be too expensive, a stair lift is one alternative. Lifts should have acoustic and visual signals indicating on which floor they have stopped. The control panel should indicate the floor number in relief lettering. A wheelchair lift can be an effective aid as a supplement to stairs, when it involves movement over minor level deviations.

Figure 6.9   Lifts (elevators) for workers with disabilities

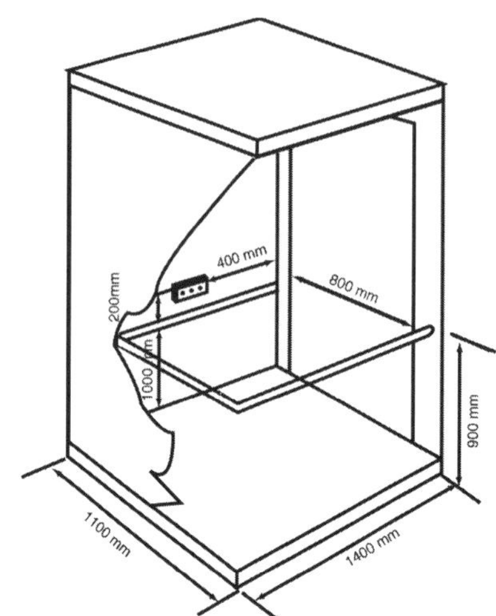

Source: Elmfeldt et al., 1983.

Interior stairways

Keeping in mind the needs of workers with a mobility impairment, as well as wheelchair-confined persons, stairs should be one of two connecting means between two floor levels. The stairways shall only be a complement to lifts or, in certain cases, ramps. In order to reduce risks to visually impaired workers or those with movement difficulty, isolated single steps should be avoided.

Ramps

Surface level differences that cannot be handled by lifts, platform lifts or the like, should be eliminated to the extent possible. In order to negotiate the smaller level differences, however, ramps can be used. These ramps should be as level as possible.

Doors

Entryways, doors to lifts and corridor doors require an unrestricted minimum breadth of 800 mm.

In order to open and close a door from a wheelchair position, manoeuvring space surrounding the door is required. The minimum unrestricted space for a common-hinged door is 500 mm. The minimum manoeuvring space surrounding entryways, lift and corridor doors is 700 mm.

In order to reduce the risks of accident for blind and visually impaired people, the doors should be braced, when open, against the wall. Likewise, doors opening onto the corridors and other trafficked rooms should be eliminated. On the other hand, doors to lavatories can open outwardly so as not to utilize the floor space within the lavatory.

Regular door handles and lock fittings can generally be used in order to open, close and lock a door. In spaces where wheelchair users must close the door behind themselves, for example in a lavatory, a horizontal draw-handle bar should be placed on the door's surface about 800 mm above the floor (figure 6.10).

Figure 6.10   Doors for workers with disabilities

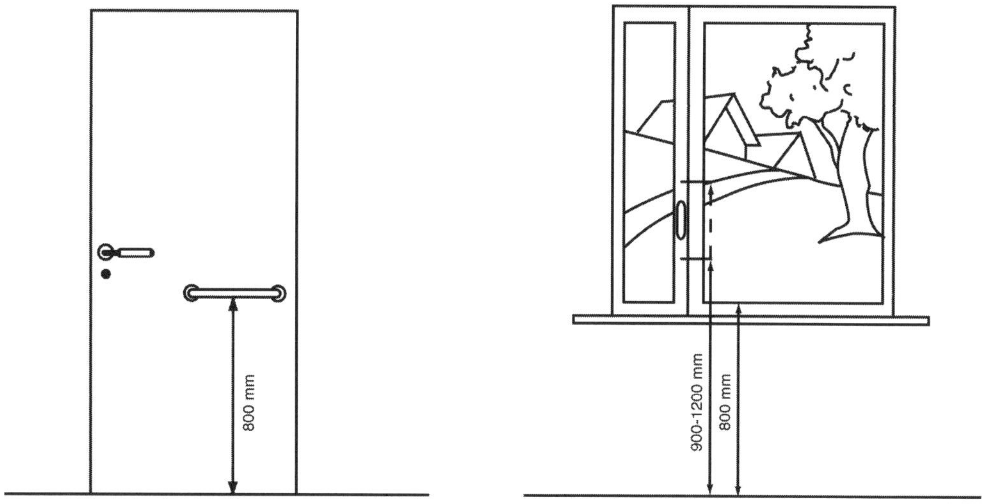

Source: Elmfeldt et al., 1983.

In order to facilitate the opening of doors that are equipped with a door closer, they should be fitted with a vertical easy-to-grip draw-handle. On the other side of the door, a common pressure plate can be used. Disabled workers with a mobility impairment and persons with reduced movement and strength in the arms and hands have difficulty in opening and going through doors with a pressure door closer. Such doors should be easy to open and preferably equipped with a motor-driven door opener.

Doors that are kept closed for fire safety purposes can remain open if they are held open by an electromagnet, which will release the grip and permit the doors to close in the case of smoke formation.

Doors sills should be low and preferably eliminated.

In order to facilitate orientation for visually impaired people to locate and negotiate glass doors, they can be complemented with a contrasting colour field, frames or similar design directly at eye level.

Windows

In order to make large glass panes that are adjacent to the passageway more visible, the panes should be marked with a contrasting colour field or frame directly at eye level. Keeping in mind risks associated with glare, the window should be located so that reflections in the floor and walling materials are avoided and so that direct glare from the window is avoided as well. See figure 6.10 for suggested dimensions.

Lighting

Lighting has great significance for people with varying degrees of residual vision. Through suitable lighting in combination with the correct material and colour choice, vision is facilitated.

The lighting situation is complicated by a number of factors, e.g. owing to the increased need for lighting and the intense sensitivity to glare. One must, therefore, place high requirements on light distribution, contrast and anti-glare. Light colours should generally be used.

It is important to know that the eye is sensitive to colour differentiations only when the level of brightness is sufficient. The difference in the brightness between various surfaces should not be so great that the eye tires and becomes irritated.

Control units

For wheelchair users, it is advantageous for control units and panels to be placed at a low level, while for the visually impaired, it is better that the control panel be placed at eye level. So that the wheelchair user can reach the control panel and, at the same time, not be too low for the visually impaired, the control panel should be placed in the range of 900-1200 mm above the floor and at least 400 mm from the corner (figure 6.11). The control unit should also be in contrasting colour to the background in order to facilitate its location by the visually impaired. In addition, necessary information should be presented in relief/Braille for tactile reading.

Figure 6.11   Control units for workers with disabilities

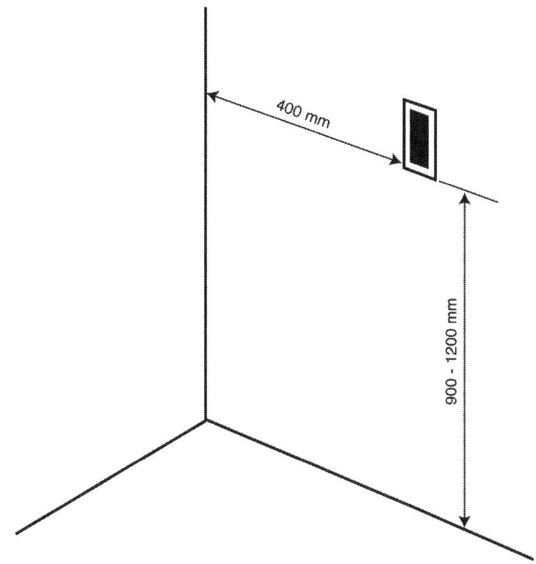

Source: Elmfeldt et al., 1983.

Handrails

A handrail can be used, in part, as support and assistance in movement by people with mobility impairments and, in part, as an orientation aid for blind and visually impaired workers. In order to facilitate orientation, the handrail should be of a contrasting colour to the background wall. With the risk of contact allergies in mind, materials containing nickel and chrome should be avoided in the handrail construction.

Signs

Signs should be designed and placed in a uniform manner, so that they are easy to read and easy to find. The text and general information should be clear and easy to understand. Signs should be placed at eye level and it is important to be able to move close to the sign so as to read it. The proper height of the text varies according to the reading distance. The height of the text, however, should never be less than 12 mm.

To enable blind and severely vision-impaired people to read tactually, texts and symbols should be presented in relief.

Signs should also be well-lit. Additionally, the sign materials should be such that they do not present a glare or reflection problem. For this reason, signs should not be placed behind glass or similar material. Information and sign displays that are only produced in slight colour difference should be avoided, as they are more difficult to see by colour-blind and visually impaired people.

For the wheelchair-confined person, it is above all important that the lavatory is suitably spacious to permit manoeuvring of the wheelchair and comfortable movement to the toilet and wash basin. Figure 6.12 shows one example of a lavatory that, in most instances, fulfils these requirements. For the wheelchair user, certain adaptations and special equipment are often also necessary.

Tile and similar material can give off reflections that are annoying to the visually impaired. By utilizing contrasting colours between, for example, the floor/wall and fixtures/walls, this problem can be reduced.

Figure 6.12　Lavatory dimensions for workers with disabilities

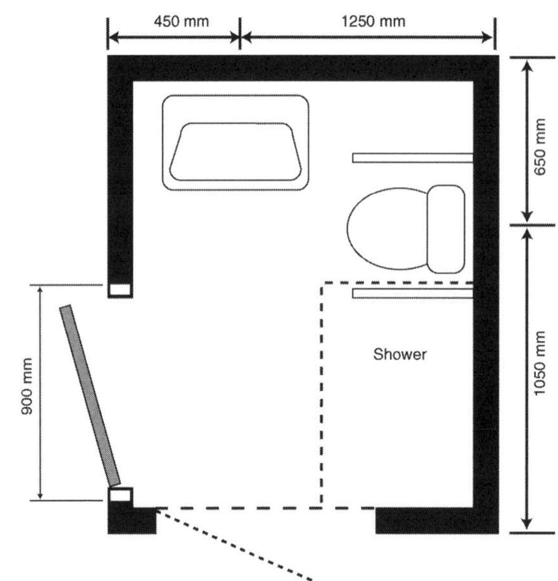

Source: Elmfeldt et al., 1983.

References

C. Benz, R. Grob and P. Haubner: *Designing VDU workplaces: Workstation, environment, organization and system introduction* (Cologne, Verlag TÜV Rheinland, 1983).

Canadian Standards Association: *A guideline on office ergonomics*, document no. CAN/CSA-Z412-M89 (Rexdale, Ontario, 1989).

E.N. Corlett, J. Manemica and R.P. Bishop: "The design of direction finding systems in buildings", in *Applied Ergonomics*, Vol. 3, No. 2, 1972, pp. 66-69.

G. Elmfeldt, C. Wise, H. Bergsten and Å. Olsson: *Adapting worksites for people with disabilities: Ideas from Sweden* (Solna, Sweden, National Institute for Working Life, 1983).

J. Henriksson and S. Lindqvist: *The apartment factory* (Örebro, Sweden, State Council for Research in Construction, 1977) (in Swedish).

—: "Buildings for work", in *Exempel 86/87: Annual review of Swedish architecture* (Stockholm, Sveriges Praktiserande Arkitekter, 1987), pp. 17-23.

B. Holdsworth and A. Sealey: *Healthy buildings: A design primer for a living environment* (London, Longman, 1992).

T. Ivergård: "The use of ergonomics in the design of new industries", in *Applied Ergonomics*, Vol. 4, No. 3, 1973, pp. 150-153.

N. Kwallek and C.M. Lewis: "Effects of environmental colour on males and females: A red or white or green office", in *Applied Ergonomics*, Vol. 21, No. 4, 1990, pp. 275-278.

B.E. Makie and G.R. Fernie: "Impact attenuation of floor coverings in simulated falling accidents", in *Applied Ergonomics*, Vol. 21, No. 2, 1990, pp. 107-114.

Nixdorf Computer AG: *Ergonomie Grundlagen der Arbeitsplatzgestaltung* (Paderborn, Germany, 1989), pp. 23-25.

Rabinda Nath Sen: "Certain ergonomic principles in the design of factories in hot climates", in ILO: *Inter-regional Tripartite Symposium on Occupational Safety, Health and Working Conditions Specifications in relation to Transfer of Technology to the Developing Countries* (Geneva, 1981), pp. 123-147.

J.E. Thurman, A.E. Louzine and K. Kogi: *Higher productivity and a better place to work: Action manual* (Geneva, ILO, 1988 and second impression with corrections, 1994).

STRATEGIC ORGANIZATIONAL ISSUES   7

David Buchanan

This chapter introduces three key strategic issues which companies concerned with work organization and ergonomics should also consider. These are topical issues, and can be considered strategic because they affect the competitive strengths of the organization. The three issues are:

- total quality management;

- just-in-time manufacturing; and

- information technology applications.

The purpose of this chapter is to introduce these issues for consideration at enterprise level – as matters of policy, to highlight the issues, the benefits and limitations, and to put these topics on the senior management agenda. There is not space to offer a detailed implementation guide under these three headings. What is provided is a briefing to enable a project team or steering group to decide whether, and how, to pursue these issues further. The nature of the implementation process involved in each area is also explained briefly.

Total quality management (TQM)

Total quality management implies a shift in management focus, which is directed towards:

- process improvement   (rather than employee control)
- prevention of defects   (rather than re-work)
- customer needs   (not company specifications)
- employee empowerment   (not managerial prerogative)
- high-involvement management   (not directive supervision)
- continuous improvement   (rather than one-off gains)
- team-based problem-solving   (not management dictation)

Most companies adopting this approach begin with a clear statement of "mission" and "core values".

The ultimate aim of a total quality approach is improved competitiveness. This can be achieved by improved customer satisfaction through increased quality of product or service, better delivery or lower price; improved internal communications and problem-solving; improved utilization of plant and facilities; improved safety; financial savings through waste reduction and other improvements; increases in innovations; and faster problem detection and resolution.

Total quality management is thus easy to explain and to justify. However, it can be less easy to implement because the approach has to be tailored to fit different organizational settings, and because it takes time to make the required organizational changes. Conceptually simple, logistically difficult would be a good summary.

A new approach to total quality

Motorola claims that total quality techniques have cut approximately US$700 million from manufacturing costs in five years. British Airways' engineering division in 1991 underspent their budget by UK£38 million, despite higher materials costs and a significant pay rise. This was attributed directly to a total quality programme designed to cut operating costs. Many companies attribute a turnaround in their fortunes to the application of total quality management principles.

However, a survey in 1992 by the consulting firm Arthur D. Little, covering 500 American manufacturing and service companies, found that only one-third felt their total quality programmes had a "significant impact" on their competitiveness. While the overall approach appears straightforward and the benefits seem clear, achieving these benefits takes time and effort, and can be problematic.

Cost and quality used to be considered mutually exclusive. If you wanted high quality, this would be reflected in the high cost of manufacture, and ultimately in the price. There are two respects in which this traditional perspective is no longer appropriate, and a fresh approach is required.

First, the concept of quality has to be defined with respect to *what the customer wants and expects*. The concept of quality, which means "fitness for purpose", is quite different from equating quality with luxury. Customer satisfaction is a goal at the heart of the total quality movement. Meeting the needs of the customer does not necessarily mean providing an extremely expensive item or service.

Second, many organizations today find that they are competing on *both price and quality*. Customers expect appropriate levels of quality, at a competitive price, with delivery according to precise requirements. This packaging of related concerns – for cost, quality and customer orientation – is sometimes described in terms of "the search for excellence". Contrary to traditional management thinking, higher quality can lead to lower cost and improved productivity: making the product or providing the service to the required level of quality first time around is less expensive than providing something defective and fixing it at a later stage.

The two routes to quality management

An organization considering the implementation of total quality management has two routes from which to choose. These can be simply described as the systems route and the culture route.

The systems route involves compliance with the International Organization for Standardization (ISO) 9000 standard. The main elements of the accreditation procedure for ISO 9000 are outlined in the following section. The emphasis in this approach is on demonstrating that the organization has appropriate and effective systems and procedures for establishing and maintaining quality of product or service, for establishing and rectifying faults, and for handling customer enquiries and complaints. Quality in this approach is achieved and maintained through systematic documentation and monitoring, reinforced by regular audits from the accrediting body.

The culture route involves establishing a corporate mission and core values which encourage a change in attitudes towards work and customers. Put simply, this means establishing a quality culture in the organization. The key values include customer care, employee empowerment and continuous improvement. The key techniques typically include a change in management style and the use of different types of problem-solving teams and task forces. Every employee in this approach delivers a product or service to customers, as many of those customers are internal to the organization. The phrase "customer orientation" thus applies throughout the organization, and not just to relationships with the "end users" of its product or service. This is sometimes expressed by the concept of a "customer-supplier chain". Quality is achieved and maintained in this approach through permanent changes in attitudes and behaviour, reinforced by changes in organizational culture and management style.

Explained in this way, these two approaches can sound distinct and exclusive. That is not the case. Many companies have used the implementation of ISO 9000 to trigger wider and longer-term changes in company culture and structure. On the other hand, companies that have sought to introduce a total quality culture have found it useful to support this change with accreditation to one of the leading standards bodies.

How to choose between the two modes? It is difficult to argue that one approach is better than the other. The appropriate choice will depend on existing organizational culture and previous history. The culture change approach may be seen as too vague and abstract in some settings, the systems approach may be seen as too mechanical and bureaucratic in others. The choice of route to follow will thus rely on an assessment of what the organization is seeking to achieve, and on what the organization is ready to accept at the point in time.

ISO 9000: Systems and procedures

The systems route to total quality means satisfying an assessor that your organization has satisfactory procedures, appropriately documented, to ensure consistent product or service quality. The focus in this approach is on systems and procedures. Specified in this way, quality management can sound like a detailed bureaucratic approach. However, the necessary systems and procedures are straightforward and are simply good management practice. Attention to these details can trigger the changes in employee attitudes and behaviours that characterize the cultural approach to total quality management explained in the next section.

An organization seeking ISO 9000 certification has to produce evidence of the following:

- an unambiguous statement of management responsibilities;
- a quality manual which defines systems and standards;
- procedures for monitoring customer satisfaction;

- procedures for monitoring design activity;

- proper storage and control of relevant documentation;

- mechanisms for the assessment of quality of incoming goods and services;

- traceability through the manufacturing process to determine accountability;

- adequate process controls;

- procedures for inspection and testing and adequate maintenance of inspection records with quality standards also for the recording system;

- procedures to ensure adequate calibration of machinery and its safe operation;

- appropriate statistical quality controls;

- fault controls, and procedures to prevent supply of faulty products;

- investigative and preventive mechanisms responsive to quality problems;

- adequate product storage and handling facilities and procedures;

- regular quality audits;

- appropriate staff training in quality systems; and

- product servicing defined and recorded.

These criteria have to be defined and applied differently, depending on the business context. They can also be interpreted to apply to service operations as well as manufacturing.

The main disadvantage of this approach lies in its potentially narrow attention to the details of paperwork and procedures. There is a danger that the paperwork takes priority, and the needs of the customers then suffer. A second disadvantage is the lack of glamour that can accompany this procedural approach to total quality. Concentration on the operational details may be appropriate for some companies, but an added danger may be that this is carried out in isolation from the development of strategy and longer-term planning.

However, ISO accreditation is a significant symbol of quality. Most companies are proud to display the symbol on their paperwork and on their products. This confers marketing advantage, providing reassurance to potential customers who may not have much prior information about the company. ISO 9000 may be a prerequisite for export into some markets. The systems and procedures are probably the same as those required to sustain a culture-based approach to total quality anyway. Many organizations have used the systems and procedures approach to trigger the longer-term change in employee attitudes and organization climate that the culture-based approaches aim for.

How to approach TQM

The culture route to total quality means changing attitudes and behaviours, at all levels, to encourage concentration on customer needs and continuous improvement. This involves changes in management style, the introduction of teamwork, a participative approach to decision-making, and the notion that everyone has a customer to service, even where those customers are inside the organization. This approach can be expressed in terms of the contrasts listed in table 7.1.

Table 7.1   Contrasts in managing for quality

Total quality is:	Total quality is not:
Meeting the requirements of internal and external customers	Expensive, luxury, top of the range
Consistent performance	Good enough
Preventing mistakes	Find it and fix it later
Prioritizing improvements	Random fire-fighting
Assessment against customer requirements	Guesswork
Everyone is responsible for quality	The quality department is responsible for quality

Most total quality manuals offer lists of "commandments" or "principles" on which to base a company-specific approach. One of the best known prescriptions is W. Edwards Deming's 14-point philosophy for management (see Hodgson, 1987). His approach in summary involves the following:

1.   *Constancy of purpose*

This refers to the clear definition of short- and long-term goals, and the devotion of resources in innovation, training, maintenance of facilities, and constant improvement of product and service design.

2.   *Adopt the new philosophy*

This refers to the philosophy of customer care (considering "the customer" as the next person down the chain after yourself) and the concern for quality defined in relation to the needs of customers. It also involves recognition of the fact that it is more expensive to fix a faulty item than it is to make it right first time around.

3.   *Do not depend on mass inspection*

Inspection, by definition, happens *after* mistakes have been made and is thus an expensive step. The existence of a separate test or inspection department is also organizationally divisive. More effective, Deming points out, is to improve the manufacturing process, and this is achieved through rigorous statistical quality control methods, and an organization which puts people in charge of what they do, including the authority to achieve quality.

4.   *Do not choose suppliers on price*

Choose suppliers that can provide consistent quality, then consider price in that context. Reduce the number of vendors, using only those whose quality meets your standards.

5. *Constantly improve manufacturing and delivery systems*

The search for continuous improvement is at the heart of the total quality approach and is thus a key management responsibility. Reducing waste and improving quality means improving productivity. Traditional management thinking claimed that a quality focus would slow down processes and reduce productivity. Deming points out that errors and faults have a more damaging effect on productivity by creating waste and re-work; get it right the first time and keep improving the process and productivity must rise.

6. *Provide on-the-job training*

Constantly improve the skills of "front-line" employees and use statistical methods to determine training and retraining needs.

7. *Redefine the role of supervision*

The role of the supervisor is to improve the system, to help people and equipment do a better job. This also means empowering first-line supervision to tell higher levels of management about problems that need to be corrected – and management must then commit itself to taking corrective action.

8. *Drive out fear*

Management by fear is a sign of weakness. People must feel that they can ask questions, report problems, challenge existing methods, and challenge management. This is central to continuous improvement in quality and productivity.

9. *Remove functional barriers*

Everyone in the organization has a customer, and everyone must be encouraged and be able to work as part of a team in the interests of the company. Internal competition and conflict between departments and divisions is wasteful.

10. *Eliminate quantifiable goals*

Numerical goals distract attention from the real problems, and can encourage harmful internal competition. The output figure may rise, but variances in raw materials may have been ignored to achieve this. The output figure may fall, and this can trigger a search for who to blame, rather than action to overcome the problem. Fixed targets, posters and slogans are not effective ways to improve productivity. A focus on working smarter and on continuous improvement are more effective.

11. *Eliminate work standards*

If you allow for 10 per cent to be defective, that can guarantee you will get that specified amount. Quotas can destroy pride in workmanship, as somebody on piecework can get paid for producing defective items. The introduction of penalties for defective work creates arguments over which cause to blame (worker, system, equipment). Numerical targets ignore the quality of product or service on which the company's reputation and goodwill depend. Deming argues that piecework and work standards are evidence of inability to understand and provide adequate supervision.

12. *Remove barriers to productivity*

These can include poorly maintained machinery and tools, poorly specified procedures, faulty inspection instruments – anything that prevents someone on the shop-floor or in the office from doing a competent job.

13. *Educate and train*

Total quality involves an understanding of the process, of statistical quality control methods, teamwork, and team problem-solving and decision-making. Employees at all levels need these skills.

14. *Top management drive*

The management structure has to drive the other 13 points daily and consistently.

This approach requires a fundamental shift in organizational culture, affecting roles, attitudes and behaviour at all levels, from senior management down to the shop-floor or in the office. That shift in culture is broadly consistent with research and commentary on the characteristics of the effective organization, with high-involvement management, empowered employees, and a clear customer-care orientation. The benefits from this change in culture can be highly significant. The main difficulty with the approach is that it is specified in general terms and has then to be tailored for implementation in particular company settings. It can be difficult to decide when and how to begin.

Implementing TQM

In some respects, the implementation of a total quality management approach is simple. It means taking appropriate action to make sure that the company's products and services do indeed conform to customer requirements. The logistics of achieving this simple goal can, however, be extremely difficult to establish. Many companies embarking on this route are likely to find that they do not have a clear understanding of customer needs, and the first step may be to establish these and to identify clear product specifications and levels of service. Often the action required will cover a range of issues and areas, and it is necessary to establish priorities before proceeding. Consistency of approach is more easily implemented in a small organization, and is considerably more difficult in a multi-site operation. Most companies will be looking for early successes from a total quality management programme. While these may be forthcoming, it is necessary to recognize that the overall time-scale for implementation – and to see significant and sustained results – can be two to five years.

To complicate matters further, every company is unique, with its own culture, values and philosophy. An approach to total quality management has to be tailored accordingly. The experience of other companies with the approach can only be a rough guide or a source of ideas and inspiration, not a template to follow.

The company choosing total quality management is likely to go through the following steps:

1.   run a senior management workshop, to establish a common strategy and a commitment to the approach;

2.   conduct an employee attitude survey, with an analysis of quality issues;

3.   identify and prioritize areas for action in terms of where the most significant results can be achieved;

4.   establish task forces and identify task force leaders to tackle the priority issues;

5.   ensure that success is publicized and recognized;

6.   arrange employee briefings on total quality management principles, and explain the role of improvement groups;

7.   set up improvement groups or quality circles and train group leaders in teamworking, group problem-solving and decision-making techniques;

8.   establish a steering committee on quality issues, with top management support.

A task force is a small group concentrating on a specific problem issue, with the problem and membership decided by senior management or a management steering committee. Improvement groups, on the other hand, are smaller volunteer teams, typically from the same organizational level, who have decided to address problems that they have identified for themselves.

The key success factors are well known. These include:

*   *sustained top management commitment;*

*   *employee awareness and understanding;*

*   *total quality skills training.*

If senior management do not have and visibly display long-term commitment to a total quality approach, the implementation is likely to fail. If employees at all levels do not share a common understanding of the total quality approach and what it means for them, the implementation is likely to fail. If employees at all levels are not given appropriate and effective training in total quality principles and in the support techniques of teamwork, problem-solving and decision-making, the implementation is likely to fail.

It may be useful to appoint a full-time facilitator to assist the implementation process. It is also helpful to give consideration to the vehicles for internal publicity for the programme, to keep everyone in the organization informed of developments, successes and future plans.

Most organizations employ an experienced outside consultant to help drive the implementation process, particularly in its early stages. In choosing a consultant, sensitivity to the need to tailor the approach to the particular organization's current culture and needs is critical. There is now available a large number of manuals, guides, training packs, training programmes and video support materials to introduce total quality management principles and methods. New materials in this field are appearing constantly. These can be used as part of in-company briefing and training programmes.

Just-in-time (JIT)

Just-in-time methods are often introduced as part of organizational packages which include total quality management, human-centred manufacturing, and innovative production techniques. The reader should therefore refer to other relevant parts of this chapter as well as to other chapters, particularly Chapters 4 and 5, of this book.

Reference to Chapter 5 will show that the purposes of autonomous working groups outlined there, and the implementation of a TQM approach, have overlapping purposes and very similar directions for their introduction and operation.

Reference to Kanban and tooling-based solutions to improving production flow will provide practical information on some techniques to develop and practice JIT.

Put simply, just-in-time methods seek to simplify the manufacturing process. As one commentator defines it (Oliver, 1991, p.19):

> "... a just-in-time system means simply that final assembly produces goods just-in-time to be sold; subassemblies produce goods just-in-time for final assembly; and bought out parts arrive from outside suppliers just-in-time to be fabricated into subassemblies".

So, as with total quality, the manufacturing operation at all stages is geared towards the precise needs of the market-place and the customer. Also as with total quality, this means scheduling activity with respect to the needs of internal as well as external customers. One immediate consequence of this approach is the elimination or drastic reduction in inventory held as work-in-progress or as buffer stocks in the manufacturing process. This reduction in inventory can reduce materials holding costs spectacularly, releasing in some cases very significant amounts of working capital. Most of the other consequences and benefits of a just-in-time approach flow from this inventory reduction. One immediate benefit flows from the reduced floor space requirements.

The just-in-time approach requires a different way of thinking about slack in a production operation. Traditional manufacturing operations focus on labour and machine utilization. If people and equipment are not being fully used throughout the shift or the working day, then there must be a problem. Return on investment would surely be higher if people and machines were working for a higher proportion of the scheduled working time. Just-in-time thinking says the opposite, for three reasons. First, a system so fully loaded

Box 7.1   Basic advantages of just-in-time manufacture

The elimination of waste

- **Reductions in inventory levels**

- **Gradual elimination of work-in-progress**

- **Improved quality (and therefore reduced scrap)**

- **Reduced tooling costs**

- **Better utilization of floor space**

Improvements in manufacturing

- **Fast response to changes in product mix/volume**

- **Greater manufacturing flexibility**

- **Reduction in product lead-times**

has little or no scope for flexibility of response to changes in customer or market demand. The absence of slack can be uncompetitive. Second, a system that concentrates on asset utilization in this way is going to be making products that will sit idle in buffer stocks or in warehouses. Materials storage is expensive, so why tie up working capital in this way? Third, buffer or insurance stocks can act as a cover for inefficiencies which are then not effectively addressed because there is no pressure to do so.

The employee at machine or process A only produces output as long as machine or process B (the next internal customer in the chain) needs it. If process B has enough material to work with, then process A sits idle. The employee at process A can, of course, carry out basic maintenance and area cleaning tasks in such circumstances. Without careful training and briefing, however, employees working in such an environment can suffer anxiety and insecurity when they can see no obvious backlog of work for them to do.

Just-in-time manufacturing addresses these problems by introducing high levels of discretion for shop-floor staff. Just-in-time methods can be applied effectively in cells or with group technology, and this can encourage the introduction of self-managing, autonomous teams. Because of these implications for management style, employee autonomy and teamwork, just-in-time has been publicized as an approach to the improvement of quality of working life.

However, the just-in-time approach to work-flow management has limitations:

- It is more suitable for products which face a stable demand for a stable design, and is less suitable in more changeable product market contexts.

- A manufacturing system that has such tight internal interdependencies is highly vulnerable to disruption from many sources, such as employee absence, materials shortage, equipment and process failure, and so on.

- The lack of insurance stocks and the urgency for continuous improvements increases the pace and intensification of work, and can increase levels of stress among shop-floor employees.

- The just-in-time discipline is typically extended to component and materials suppliers, who must themselves then introduce the same internal procedures in their operations.

Implementing just-in-time

As with total quality management, the implementation of just-in-time methods implies a new approach to management, a new philosophy, and a new perspective on the manufacturing operation. Without these shifts in management thinking, the approach is not likely to be effective.

The implementation of just-in-time can take several months or even years to achieve, depending on the size and complexity of the manufacturing operation. Typical stages in the implementation process are:

1. establish a steering group or project team to explore the benefits and implications of a just-in-time approach for the company, and to draw up an implementation plan;

2. identify key business goals to be achieved through implementing just-in-time methods;

3. arrange study visits to comparable reference sites to learn from their experience;

4. formulate a specific, phased implementation plan, starting with one key stage of the manufacturing process and diffusing gradually through the whole operation;

5. launch an education and training programme covering all levels of staff affected directly and indirectly by the just-in-time approach;

6. encourage suppliers to arrange just-in-time delivery of supplies, thus reducing order levels and costs, and also reducing the storage space requirements for incoming goods which may be shipped straight to the manufacturing operation.

The following account of JIT implementation is a typical example of the approach in use in a medium-sized manufacturing operation.

CASE-STUDY 7A

Just-in-time: The philosophy of working properly

The attitude of McDonnell Douglas's computer manufacturing subsidiary in the United Kingdom to the Japanese production system was reported as follows:

The Hemel plant, which manufactures super-minicomputers, is a useful case-study in how to install JIT into a middle-sized manufacturing plant. Introduced over a three-year period, JIT has saved one-third of factory floor space and has allowed the company to maintain its inventory at about UK£7.5m while increasing throughput at a compound rate of 20 per cent a year. The computer system business at Hemel, in which McDonnell plans soon to reduce its stake to a minority, employs 150, and had sales last year of UK£131m on which it made a pre-tax profit of UK£21.5m. The non-union plant makes just 600 computers a year, so it is not a high-volume operation. However, its managers believe the steps they took to introduce JIT could be used as a role model for similar sized companies.

- The first phase was to make a close examination of a JIT plant making similar products. Hemel managers had the benefit of studying a sister plant within McDonnell Douglas in California where the introduction of JIT had been by no means totally successful.

- A team was set up at Hemel to study JIT further and make recommendations. It included people from all sections of manufacturing, including materials control, production control, production engineering, computer systems and accounting. The initial phase took nine months.

- During this period the fitting area – the middle of the process of making a super-minicomputer – was chosen as the place to start JIT. This was because jobs in fitting are measured in hours rather than days. If mistakes were made in the JIT system, these could be overcome by working more overtime.

- While JIT was being examined – which also involved visits to other JIT plants by managers and production workers – a large education programme was started. All employees received information on the overall aims of JIT at Hemel and on how systems worked at other companies. It was also realized that the changes would require shop-floor workers to do a wider range of jobs. This was because they would occasionally be transferred to other parts of the factory which needed more hands for the tight JIT-governed production flow. A table of all production workers and their skills was drawn up. This showed that 30 per cent of production employees could do 70 per cent of the jobs. A big retraining programme to give them extra skills was started. At first, Hemel thought it could do all this itself but found it could not. The Production Engineering Research Association (PERA) at Melton Mowbray was called in at a cost of UK£10,000; its staff helped with retraining and also trained people to continue with it once PERA had left. Laver says particular attention was paid to foremen and supervisors who would no longer have a traditional role. Because workers would be switched from time to time to areas traditionally commanding higher – and in some cases lower – pay than those they were used to, everybody went on to the higher rate of shop-floor pay. Laver says it would have been folly to do otherwise. So that workers could participate in the process, they were given the opportunity to design their own JIT system. The company found, though, that its professional engineers ultimately provided the solutions.

- JIT was introduced to further parts of the factory over the following two years, finishing up with the taking in of components and the despatch of finished goods. These later phases of JIT were started off with just 25 per cent of "slack" in trigger levels. In the past few months, staff at Hemel have been attempting to spread the message to their suppliers.

Source: Garnett, 1989.

Technological change

Computing technologies and information systems still continue to develop and to penetrate working life in many forms. New technology encourages changes in job design and organizational structure, and can open up a range of new individual and organizational opportunities. However, new technology can also create training, redeployment and redundancy problems. The implementation of new technology can also be awkward and problematic if not managed professionally. The opportunities and the problems are organizational as well as technological.

The key questions which the following sections will help to answer are:

1.   What new technology is appropriate for the organization?

2.   What business objectives will it help to achieve?

3.   What changes in work organization are required, to ensure quality of working life and effective use of the new technology?

4.   How will the technical and organizational changes be implemented?

The focus of the section will be on the third and fourth issues, concerning work organization and implementation. We can offer only general guidance in the first two areas, for two reasons. First, the range of applications is now too wide to cover in one book; there are many specialist publications available which meet this need. Second, there is variation in technology and business objectives from company to company and the answers to those first two questions are site-specific.

This issue of technological change is important for a number of reasons.

Continuing innovation

One reason is that computer-based technologies and information systems continue to develop. Many applications are today replacement or update applications, and not first-time adoptions, and new issues are raised in this update and extension process. The objectives behind replacement applications are often quite different from those which prompted initial adoption. The experience and skills of employees with respect to computerized systems are likely to be much more extensive. The implications of technological change in this context are thus different.

Organizational issues

A second reason concerns the evidence which shows that many (if not most) application failures and problems are caused by organizational issues, and not by technical factors. Research findings consistently reinforce this picture. R. Long argues from his research that most problems in office automation applications are due to poor planning, poor technical management skills, lack of user training, and uncertainty about the "right" problems to address.

A survey of 400 British and Irish companies, carried out in 1990 by consultants A.T. Kearney, revealed that only 11 per cent had been successful in their applications of information technology, on criteria concerning breadth of applications and benefits achieved, project completion on time, and return on investment. The consultants' report concluded that this rate "must be judged unacceptably low", and also indicated that the reality may be bleaker since this was a self-selected set of respondents to a mail questionnaire which was presumably returned by the more competent and, in their own estimation, effective companies. Among the recommendations for management action from this report were the following:

– recognize that organization and people issues, not technical areas, are the barriers to success;

– understand that organization and people issues are also the key contributors to success;

– change the organization structure to capitalize on the strategic competitive advantages offered by information technology; and

– appoint a user manager as project leader, not a computing or data-processing specialist.

Successful applications are dependent on effective implementation – not just on good system design.

Design space

One view of new technology application is that the technology determines the design of tasks and roles necessary to operate it effectively. The evidence suggests that this technologically determinist perspective is inappropriate. Instead, it seems that in most situations, the introduction of new technology opens up a window of opportunity with respect to ergonomics and work design. This window has also been called design space, to indicate the existence of a range of work and organization design options.

To be able to identify and to assess such options and to make the best decisions when introducing new technologies, it is in turn important to involve the people concerned. Participative approaches thus become an essential element in technology implementation. The criteria for choice within the design space are also potentially negotiable, and can include issues of health and safety, quality of work life, operating efficiency, and ease of maintenance and repair.

The implementation of new technology, and the consequences of change, thus rely on an organization decision process, and not just on the capabilities of the new equipment and systems.

Box 7.2 summarizes the main conclusions of a series of demonstrator projects which evaluated applications of office automation systems in a range of different organizational contexts. The project was sponsored in the mid-1980s by the Department of Trade and Industry in the United Kingdom. This experience produced a number of key messages with respect to new technology in general, and to office applications in particular.

Box 7.2 Key messages for new technology introduction

- Do not rush into a project – know where you are starting from and plan.

- Take a phased approach and spread by successful example, starting with critical applications and demonstrable benefits.

- Identify clearly the benefits to be sought and actively manage to achieve them.

- Office automation needs technical expertise, but it must service and support applications, not drive them.

- People are the most important resource – communicate early and often, get them to contribute, and deal positively with natural fears and concerns.

- Planning is important, but so are leadership and management, and allowing users to generate development through creative freedom.

- Systems are rarely turnkey – buy-off-the-shelf, but plan to tailor and develop.

- Integrate the technology into the organization, into jobs, onto working procedures, into career paths, into management goals.

Source: United Kingdom. Department of Trade and Industry, 1985.

The key components of an organizational decision process are summarized in figure 7.1, which illustrates the main dimensions of the process determining the consequences of new technology application. The outcomes of technical change subsequently influence future technology investment decisions, influence future management objectives with respect to innovation, and influence work organization decisions as a consequence of psychological assumptions made about people (concerning the relationship between their skills and involvement and their motivation and cooperative attitudes).

Figure 7.1 Technical change: An organization's decision process

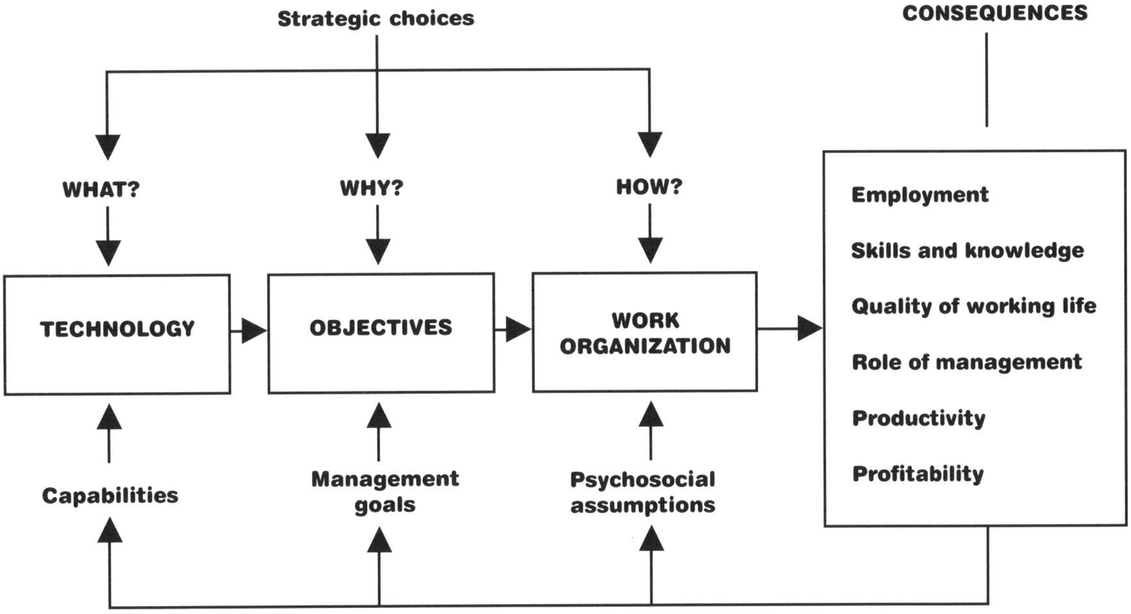

On comparing this figure with the structure of this book, you will see that certain chapters contribute in the various boxes in the figure. Chapter 4 is important to technology, while Chapters 1, 3 and 5 relate strongly to the third box (work organization) and to several of the consequences at the end of the diagram. None is exclusive to a single box; as was said early in this book, and is reinforced at intervals, the factors relating to effective and acceptable work are complex and interacting. The organization's decision processes to create effective work are iterative, and require the continuous surveillance of all the relevant factors to bring them together in the best possible way. It is a working team which brings the skills for doing this together, and itself is the nucleus for the growth of teamworking in the organization.

The technology choice: Must it always be high tech?

There is evidence that even in industrialized economies, "leading-edge" technology is not necessarily or always "appropriate technology". Some "high-tech" organizations today use comparatively "low-tech" approaches to manufacturing, involving significant human contribution, because this has proved to be more reliable than more highly automated – and considerably more expensive – manufacturing systems.

Leading-edge systems can often be expensive and difficult to justify on cost grounds. Once purchased, they can put the company on a committed path to further investment in technology, very difficult and expensive to reverse. New systems may not be error-proof, the manufacturer may be using early purchases to test out its new model, and they may be untested in particular application domains. Of course, being new technology, they will be unfamiliar to users, perhaps not just to manipulate but to understand and recognize the characteristics of the process, and they will also be unfamiliar to maintenance and engineering staff.

Leading-edge systems can therefore be risky. When highly automated systems start to go wrong, a lot of faulty items can be produced and much material wasted before anyone notices what is happening and corrects it. Affluent organizations that want to be seen by their customers as being at the leading edge can often afford to experiment with new systems. Less affluent companies whose business does not depend on technical wizardry may not find this strategy appropriate. The concept of appropriate technology should not now be seen as one applying solely in developing economies; it is a concept whose relevance depends on the nature of the business in which the organization is operating.

Before embarking on investment in leading-edge or state-of-the-art technology, it is therefore useful to consider questions such as:

- Can we afford the loss if the system turns out to be inappropriate for us?

- Do our people have the skills to operate this?

- Do our maintenance staff have appropriate skill and understanding?

- What training will we have to offer users, operators and maintenance?

- Will this commit us to a particular future technology path, in terms of further developments and upgrades, and what will this mean?

- Will this technology really allow us to provide a better product and/or service to our customers?

- Do we need to be seen by our market as a high-tech company, or is it our products and services that count?

- Will this technology help us to develop and introduce new products and services, or will it commit us to a particular product range until we reinvest?

If the company does not have the operating and maintenance skills, if the training is time-consuming and expensive, if the financial commitment is risky, if customers will not benefit directly, and if the technology is relatively inflexible, then the leading-edge technological solution may not be appropriate and should be abandoned. For further discussion, refer back to Chapter 4 and case-studies 4E and 4F.

Proven technology that is able to do the job that is required of it is appropriate and effective in many settings. Even the leading-edge companies with high-tech systems typically have conventional equipment operating alongside new systems as a fall-back, and often because there are some operations that the conventional systems can carry out better, cheaper, faster, and to just as good quality. New is not always best.

Investment in state-of-the-art systems is more likely to be appropriate:

- where the organization can afford to experiment and write off failures and mistakes as experience;

- where the organization has skilled systems people who know what they are doing;

- where the organization trades in a market that expects leading companies to use leading-edge technology;

- where the new technology will give a clear technological or/and commercial advantage.

Case-study 7B describes work by Ingersoll Engineers in which advanced, but proven, manufacturing technology achieved significant benefits. Some of the world's leading companies have rejected expensive automated systems in favour of manual manufacturing, combined with the work organization techniques described in Chapters 3 and 5. Digital Equipment in Scotland use manual assembly to build their small business computers, and rejected an expensive, inflexible automated assembly system in the mid-1980s. IBM, at their plant in Havant in England, use manual assembly of computer disks because the automated system, with robots, could not guarantee the attention to detail, product cleanliness and quality required. Once again, the capital-intensive automated solution was rejected in favour of a more conventional manual assembly approach.

CASE-STUDY 7B

Technology in manufacturing

Ingersoll Engineers sought to establish the success and failure factors in applications of advanced manufacturing technology. Their research showed that there were three main criteria for success:

1. The technology should be part of an overall plan or strategy to serve business needs.

2. The attempted leap in technology must not be too great.

3. The people concerned must be closely involved.

They found that the main single barrier to success was the failure to involve people. "Wherever we've seen manufacturing technology achieving its full capability, the people on the shop-floor feel a measure of ownership in it, and union officials feel deeply involved too. In every case of failure we've seen, these non-manufacturing people issues were not tackled vigorously enough."

The research also identified the main reasons for failure in new technology applications:

1. superficial decision-making;

2. going straight to the solution, then identifying the problem;

3. being convinced by persuasive sales people;

4. investing for today's products and not for tomorrow's;

5. wanting to belong to the prestigious "high-tech" club;

6. change by top management edict: "We'd better get it, the boss wants it".

Ingersoll report an application of technical change in manufacturing that achieved an annual increase in profitability of between US$9 and US$12 million without costing the company anything. How was this done? Here are the results.

Area of benefit	Before	After
Throughput time (responsiveness)	25 days	2 days
Additional sales		unmeasurable: "a lot"
Inventory turns per annum	5	30
Delayed deliveries	40%	2%
Fork-lift trucks	24	5
Inventory	US$10 million	US$2 million
Hours (everybody) per unit	330	200
Unexplained cost reductions	US$1 million	
Re-work (plant and field)	6%	1%

The annual value of these improvements to the company was between US$9 and US$12 million, and the cost had been US$6.5 million. In balance sheet terms, the cost was *minus* US$1.5 million because the one-time inventory reduction of US$8 million more than paid for the whole project.

.../...

Case-study 7B (cont.)

The technology used was not particularly "high". There was no flexible manufacturing system and no manufacturing resource planning. Some of the previous computerized manufacturing support was removed. There were a few robots and some direct numerical control on machine tools. No computer-integrated manufacturing and only some automation was involved. It was "low technology" that produced these stunning results. The Ingersoll research showed that this company had several features in common with many other successful companies that had achieved the same benefits:

- the technology was proven;

- the company was not held back by traditional investment justification procedures;

- there were no rigid division (or battle lines) between organizational functions;

- the company had some idea what they wanted to be doing several years into the future;

- all those who could "make or break" the project were involved enthusiastically with it;

- each application had a "champion" who sold the idea to those concerned.

Source: Ingersoll Engineers Ltd., 1987.

Management objectives and technological change

As indicated earlier in figure 7.1, technological change can be described as an organization decision process. One of the main elements in this decision process concerns management objectives.

Three categories of objectives can be identified: strategic, operational, and control objectives.

Strategic objectives. These relate to goals, such as the desire to reduce production costs, replace obsolescent equipment, overcome bottlenecks, control energy use, cut other plant or building running costs, and reduce the numbers and costs of support staff.

Operational objectives. These deal with the operation of the company, its output, quality, future products and how the effective day-to-day running is achieved.

Control objectives. These concern management goals to reduce human intervention; to replace people with machines; to reduce dependence on human control of equipment and processes; to reduce uncertainty, and increase reliability and consistency and predictability in manufacturing operations; and to increase the amount of performance information and the speed at which it becomes available.

If your organization is considering investment in new technology, in any form, it can be an interesting exercise to list the objectives for the application under three headings:

- What strategic goals are your company pursuing?

- What are the operational objectives of your company?

- What control objectives are important for the company?

As we shall see in the following section, control objectives can often be counter-productive. The evidence typically shows that sophisticated, clever, flexible, expensive systems need sophisticated, clever, flexible, expensive people to operate and to maintain them. There are, of course, exceptions to this rule, but these seem to be rare.

The evidence shows that, while operational objectives may remain important in many applications, new technology has strategic potential. This means simply that it can offer better ways of conducting existing business and ways of conducting new types of business. In other words, if used effectively, it can contribute significantly to the competitive advantage of an organization.

A main consequence of this shift in emphasis to strategic applications is that advanced technology can no longer be seen just as a support activity, automating production operations and serving some planning, control and administrative needs. New system applications need to be linked closely to company strategy. One approach to the identification of an effective company technology strategy is to link critical business success factors with potential applications. Table 7.2 gives some typical examples.

Note that in many of the examples offered here, it is suggested that competitive advantage may come to depend on superior information or knowledge, and not just on conventional criteria such as product quality or cost.

Table 7.2 Applications of information technology relevant to business success

Critical business success factor	Appropriate applications
Create new markets	– customer-supplier information systems – computer-aided new product design
Find new markets	– new products database
Automate production	– process and equipment controls – robotic assembly – programmable conveyors – computer-integrated manufacturing
Profitability	– cash flow controls – electronic point-of-sale systems
Communications	– local area networks – electronic mail – corporate telecommunications
Improve image	– word processing – desktop publishing – financial control systems
Improve control	– manufacturing support systems – intelligent process controls – automated storage – stock and production control systems
Improve quality	– quality assurance documentation – computer-assisted measuring and testing – computerized visual inspection – laboratory automation

A critical success factors approach can be used to ensure that technology strategy is consistent with company strategy. This approach also encourages a long-term view. This is important because, for many new technology investment decisions, conventional accounting procedures which seek evidence of payback over relatively short time-scales – often two years – are inappropriate. Strategic benefits are often qualitative and difficult to measure in the short term, and a long-term view is then necessary.

We have emphasized the relative importance of strategic objectives in this section. This does not deny the significance of operational objectives, or of some control objectives (such as improving the availability of information). Some applications may have very significant operational benefits and little, if any, strategic impact. Most applications will, of course, involve a combination of objectives. We wish simply to reinforce two points:

- be clear about the objectives of investment in new technology;

- explore and do not overlook the strategic potential.

Work organization and human-centred technology

The choice of forms of work organization that may accompany a given technology is wide. The introduction of new technology can trigger a work organization programme, using the techniques and approaches described in Chapters 3 and 5. Very often, the preoccupation with technical issues, and the perceived importance of control objectives, can mean that the work organization options are not evaluated, or that "scientific management" solutions are adopted uncritically. The problem is that discretion, skill and motivation go hand-in-hand. The pursuit of control objectives can lead to the design of forms of work organization that eliminate discretion and thus give users neither the skills nor the motivation to perform their functions effectively.

On the contrary, as mentioned earlier, the evidence shows that, as a general rule (and there are exceptions), sophisticated, clever, flexible and expensive equipment needs sophisticated, clever, flexible and expensive people to maintain it and keep it running smoothly. The implications for human resource management in general, and for ergonomics and work organization in particular, are therefore wide reaching. Most new technology applications have been accompanied by retraining as well as redeployment, and by skills upgrading rather than deskilling. This has been the major trend with respect to both manufacturing and office applications.

More recently, research has highlighted the fact that new technology applications encourage not only skills upgrading, but also the further extension and application of self-managing teamwork. The material and considerations introduced in Chapter 5 are relevant here also. The implications of this finding affect a number of general management and human resource management practices. Skills upgrading of the workforce can affect training and retraining policies and costs, payment systems, the role of supervision, promotion opportunities, and the wider role of management.

Another significant trend concerns the development of human-centred systems. The original thrust to reduce and minimize human intervention in work processes through automation and computerization is being replaced by a move to complement human skills and knowledge. While previous technologies sought to "automate" work processes, computer-based systems "informate", and human intervention and decision-making becomes more critical in this new context.

The effective and safe operation of new manufacturing technologies thus requires careful attention to work organization, to achieve complementarity between human and equipment capabilities, and to avoid distancing the employee from the production process. It has been demonstrated that even precision machining with CNC machine tools requires close attention to the details of the production process, and frequent manual intervention to control and correct it. Because the internal mechanisms of computerized machine tools are invisible, this places a higher premium on skilled and motivated human intervention, given the high price of equipment failure and processing error. As one commentator has pointed out (Noble, 1979, p.44):

> "What will a machine operator, 'skilled' or 'unskilled', do when he sees a $250,000 milling machine heading for a smash-up? He could run to the machine and press the panic button, retracting the workpiece from the cutter or shutting the whole thing down, or he could remain seated and think to himself, 'Oh look, no work tomorrow'".

Case-study 7C highlights the close linkage between organizational decisions, technological innovations, and people and skills at the workplace.

CASE-STUDY 7C

New technology in biscuit making

The experience of automation in biscuit making illustrates the skills implications of new technology in manufacturing. David Buchanan and David Boddy compared the implications of computerization on two occupations – doughmaker and oven operator – in a Glasgow biscuit factory. The company's first production computer was installed in the early 1970s to control the dough mixing process. To change the recipe, it became necessary to prepare a new paper tape (as with a CNC machine tool) which required computer programming skills. In the early 1980s, a new "recipe desk" was introduced to carry out the same functions, and the ingredient mix could now be changed using small thumb-wheels on the desk without special computing skill.

The job of the doughmaker who supervised the mixing process became repetitive, with a cycle time of around 20 minutes, and was classed as semi-skilled. This job used to be done by a skilled master baker, but the computer now controlled the mixing process doing 60 mixes every 24 hours. Bored doughmakers sometimes forgot to add sundry ingredients such as salt, and this was only discovered at the end of the mix or during an oven test. Buchanan and Boddy argue that, in replacing the craft skills of the doughmakers, and in requiring continued human intervention at that stage of the process, computerization had created a *distanced* role in which:

1. the process operators had limited understanding of the process and equipment;

2. the process operators could not identify the causes of equipment faults, there were no backup systems for them to operate, and specialist maintenance staff were needed;

3. the process operators became bored, apathetic and careless, and rejected responsibility for breakdowns;

4. they developed no skills which could make them eligible for promotion.

These four *distancing* features are typical of many jobs in "nearly automated" manufacturing systems, where the operator develops neither the ability nor the motivation to carry out residual functions – what the equipment cannot do – effectively.

.../...

Case-study 7C (cont.)

In contrast, the implications of computerization for the oven operator was quite different. The oven operator on each biscuit production line was responsible for baking biscuits that had the correct bulk, weight, moisture content, colour, shape and taste. This was complex, because action to correct a deviation on one of these features could affect the others. A microprocessor-controlled check weigher was installed to replace the old electro-mechanical system. As each packet passed over the weigh cell, its weight was recorded by the computer and was displayed on a panel near the wrapping machine. The computer also gave summary information on packet weights to the oven operator through a video display unit. This display was updated every two minutes and was presented in graph and digital form, also producing management reports.

The new system gave the oven operator information about the performance of the line, enabling him to make adjustments to the oven controls and to reduce waste by producing more accurate packet weights. If the packet weights were too high, and the wrapping machine could not compensate, the oven operator could arrange to adjust the weight of the dough blanks (uncooked), or he/she could increase the oven temperature to increase the bulk and reduce the number of biscuits per packet. The information from the system showed that something was wrong, but did not show what was causing the problem or what action to take to correct it. The oven operator had to take into account the properties of the flour being used and the dough that it made. When the packet weights wandered, the oven operator decided what to do to correct it.

With computerization, the oven operator had become a "process supervisor". Buchanan and Boddy argue that this new technology *complemented* the skill and knowledge of the oven operator and created a role in which he:

1.	got rapid feedback on performance and had discretion to monitor and control the process more effectively;

2.	had good understanding of the relationships between process stages;

3.	had a visible goal that could be influenced;

4.	felt that the job had more interest and challenge.

The *design space* was not explored and fully exploited in this case, particularly with respect to the doughmakers. For example, management could have:

■	designed the equipment to let doughmakers see and hear the mixing process;

■	given the doughmakers the recipe desk to adjust quantities themselves;

■	established autonomous work groups each collectively responsible for a biscuit production line.

The implications of computerization for these two occupations can thus be seen to have been shaped as much by management decision as by the technology.

Source: Buchanan and Boddy, 1983.

New technology makes employee skill and commitment *more* important, not less. Box 7.3 summarizes the main distinctions between distancing and complementing strategies for work organization.

Based on analyses and examples of this type, many companies have sought to enhance the capabilities of their employees and improve labour relations.

Walton and Susman (1987) argue that an appropriate management response should include four key ingredients:

1. First, the development of a highly skilled, flexible, committed and coordinated workforce.

2. Second, a lean, flat, flexible, innovative management structure.

3. Third, ability to retain skilled and motivated people.

4. Fourth, a strong partnership between management and trade unions.

In addition, effective management strategies, or "people policies", should include:

- job enrichment;

- multi-skilling;

- teamwork;

- skills-based payment systems;

- pushing decision-making down the hierarchy;

- supportive selection, training and management development procedures.

Box 7.3   Consequences of distancing and complementarity

Skills *distancing* leads to:

- **Boredom with work**
- **Inattention to events and details mistakes**
- **Denial of responsibility**
- **Limited process understanding**
- **Inability to make minor adjustments**
- **Poor coping with the unexpected**

Skills *complementarity* leads to:

- **Increased confidence in decisions, less guesswork**
- **Increased interest, skill and challenge**
- **Ready acceptance of responsibility to make the system work**
- **System adjustments quickly made**
- **More informed action and interpretation of events**
- **Improved anticipation of problems and emergencies**
- **Faster response to unexpected and unplanned changes**

These are precisely the issues discussed in Chapter 5. Competitive pressures are encouraging organizations to adopt these people policies. New technology also encourages these human-centred management strategies.

Implementation guidelines

Applications of new technology have characteristics which present particular challenges to those responsible for the implementation process. Typical features of new technology applications include the following:

- Applications are often expensive, and can be difficult to cost-justify using conventional accounting procedures. It is often necessary to justify investment using qualitative criteria (improved product quality, customer service), by indicating the strategic benefits in terms of sustained competitive advantage, and by taking a long-term outlook (not necessarily expecting payback within one or two years).

- The time-scale of implementation can often be long, causing others in the organization to question the benefits which do not seem to be forthcoming.

- Applications can also involve the long-term commitment of finance and other scarce resources, such as special expertise. Other projects and priorities can be competing for these resources, making implementation difficult.

- For the above reasons, new technology applications can often be seen as major risks to the organization, and this can have implications for how the technology is implemented, and the expectations that it creates.

- The organizational changes triggered by technical change can cut across traditional job and functional boundaries, threatening entrenched positions and vested interests, and generating resistance.

As we have emphasized throughout this chapter, the effective management of change means finding the right combination of technical and organizational innovation. The problems for those responsible for championing change involve identifying and clarifying the critical issues, evaluating possible courses of action, and making consistent decisions about factors that will contribute to a successful outcome.

Box 7.4 offers a set of nine guidelines to help in this process. These guidelines relate to the main dimensions of the organization decision process model on which this chapter has been based – the objectives or purpose of the change, the people issues, and managing the implementation process.

Box 7.4   Guidelines for exploiting the potential of new technology

Technology and objectives

1.   Ensure that the use of new technology has a clear strategic focus which targets long-term market objectives and not just current internal operating problems.

2.   Review and implement supportive policies in areas like employment conditions and investment appraisal, to encourage innovation and the strategic use of new technology.

3.   When evaluating system options, make sure that you choose kit to fit the current and anticipated needs of the business.

Technology and people

4.   Review work organization in relation to new technology, to encourage flexibility, creativity and skills development.

5.   Review management styles and working arrangements, to ensure that these are consistent with the strategic goals of the business, and with achieving the potential benefits of new technology.

6.   Design support systems which are consistent with strategic aims and which enable support staff to contribute in flexible and creative ways to the changing needs of the organization.

Technology and implementation process

7.   Establish a clear project management responsibility to guide the often protracted implementation process.

8.   Plan the nature and timing of employee involvement, to ensure that key staff establish ownership of new systems and procedures.

9.   Develop a systematic training programme to equip users at all levels with the competence, in the form of new skills, knowledge and attitudes, that they will need to exploit the potential of the new technology.

Source: Buchanan and McCalman, 1989.

Managing change

The most popular way to manage the implementation of technological and organizational change is through a project management methodology. The most effective way to manage this approach is through the formation of a project team with a clearly identified leader. This is discussed more extensively in Chapter 5 in the context of introducing teamworking. Although the context may be seen as narrower than that of this chapter, the requirements and procedures are very similar. As given in Chapter 5, a typical project life cycle has the following main stages:

1.   Develop a strategy.

2.   Confirm top-level support.

3.   Use a project management approach:
     - identify tasks
     - assign responsibilities
     - agree on deadlines
     - initiate actions
     - monitor
     - act on problems
     - close down.

4.   Communicate results.

This list of stages in the project life-cycle can be used as a preliminary planning agenda for the project team, or task force, or working party or steering group – different organizations use different terms here.

A project management method for implementing new technology is wholly consistent with the participative approach to change advocated throughout this book. The agenda of the project team must also incorporate ergonomic and organizational issues; for example, the design space must be adequately explored before final decisions are reached.

Selecting key people for introducing change

Research has also identified the skills and competences critical to the resource person responsible for introducing technical and organizational change. These are summarized in box 7.5. It is interesting to note that these are primarily management skills. It has been suggested that technical skills may be less important or even unnecessary in some settings, given the importance of management skills in change implementation. The project leader usually has access to technical expertise elsewhere in the project team.

Box 7.5   The skills of the change agent

The change agent introducing large-scale technical and organizational change requires the following 14 skills and competences:

Goals

1.   *Sensitivity* to change in key personnel, top management perceptions, and market conditions, and to the way in which these affect the goals of the project in hand.

2.   *Clarity* in specifying goals, in defining the achievable.

3.   *Flexibility* in responding to changes outside the control of the project manager – perhaps requiring major shifts in the project goals and management style – and risk-taking.

Roles

4.   *Team-building* abilities, to bring together key stakeholders and establish effective working groups, and clearly to define and delegate respective responsibilities.

5.   *Networking* skills in establishing and maintaining appropriate contacts, within and outside the organization.

6.   *Tolerance* of ambiguity to be able to function comfortably, patiently and effectively in an uncertain environment.

Communication

7.   *Communication skills* to transmit effectively to colleagues and subordinates the need for changes in project goals and in individual tasks and responsibilities.

8.   *Interpersonal skills* across the range, including selection, listening, collecting appropriate information, identifying the concerns of individuals, and managing meetings.

9.   *Personal enthusiasm* in expressing plans and ideas.

10.   Stimulating *motivation and commitment* in those involved.

Negotiation

11.   *Selling* plans and ideas to others, by creating a desirable vision of the future.

12.   *Negotiating* with key players for resources, or for changes in procedures, and resolving conflict.

Managing

13.   *Political awareness* in identifying potential coalitions, and in balancing conflicting goals and perceptions.

14.   *Influencing skills* to gain commitment to project plans and ideas from potential sceptics.

Source: Buchanan and Boddy. 1992, pp. 92-93.

Here, in summary, is a checklist of the key issues which a project team must address when planning the implementation of new technology:

1.   What systems are appropriate for our business?

2.   What are we trying to achieve? How will we evaluate success?

3.   How will work organization change? How will we identify these changes?

4.   At what stage and how will we involve users in planning the implementation process?

5.   Who will be project leader, and what will be the composition of the project team?

Clearly the way in which these issues are resolved will differ from one organization to another. Each of these five headings can be broken down into an extremely large number of subthemes and issues, many of which have been indicated in this chapter, and which again will differ from site to site. The intention here is to establish a clear starting-point – or a preliminary agenda – for the implementation project team which will elaborate these issues in the light of local needs and circumstances.

References

R. Beckhard and R. Harris: *Organization transitions: Managing complex change* (Reading, Mass., Addison Wesley, 1977).

F. Blackler and C. Brown: "Alternative models to guide the design and introduction of new information technologies into work organizations", in *Journal of Occupational Psychology*, Vol. 59, 1986, pp. 287-313.

D. Boddy and D.A. Buchanan: *The management of new technology* (Oxford, Blackwell, 1986).

—; —: *The technical change audit: Action for results* (Sheffield, Manpower Services Commission, 1987).

—; —: *Take the lead: The skills you need to manage change* (Hemel Hempstead, United Kingdom, Prentice Hall, 1992).

D.A. Buchanan: "Vulnerability and agenda: Context and process in project management", in *British Journal of Management*, 1991.

—; D. Boddy: "Advanced technology and the quality of working life: The effects of word processing on video typists", in *Journal of Occupational Psychology*, Vol. 55, No. 1, 1982, pp. 1-11.

—; —: "Advanced technology and the quality of working life: The effects of computerized controls on biscuit-making operators", in *Journal of Occupational Psychology*, Vol. 56, No. 2, 1983, pp. 109-119.

—; —: *The expertise of the change agent: Public performance and backstage activity* (Hemel Hempstead, United Kingdom, Prentice Hall, 1992), pp. 92-93.

—; J. McCalman: *High performance work systems: The Digital experience* (London, Routledge, 1989), pp. 204-206.

T. Burns and G.N. Stalker: *The management of innovation* (London, Tavistock, 1961).

R. Collard: *Total quality: Success through people* (London, Insitute for Personnel Management, 1989).

T. Cox and C. Mackay: "A transactional approach to occupational stress", in E.N. Corlett and J. Richardson (eds.): *Stress, work design and productivity* (Chichester, United Kingdom, John Wiley, 1981).

P.B. Crosby: *Quality is free* (New York, McGraw-Hill, 1978).

J. Cullen and J. Hollingham: *Implementing total quality* (Bedford, United Kingdom, IFS Publications, 1987).

M. Earl (ed.): *Information management: The strategic dimension* (Oxford, Clarendon Press, 1988).

—: *Management strategies for information technology* (Hemel Hempstead, United Kingdom, Prentice Hall, 1989).

— and D. Skyrme: "Hybrid managers: What do we know about them?", in *Oxford Institute of Information Management, Research and Discussion Papers* (Oxford, June 1990).

K. Eason: *Information technology and organizational change* (London, Taylor & Francis, 1988).

J.R.P. French, R.D. Capalan and R. van Harrison: *The mechanisms of job stress and strain* (Chichester, United Kingdom, John Wiley, 1982).

A. Friedman and D. Cornford: *Computer systems development: History, organization and implementation* (Chichester, United Kingdom, John Wiley, 1989).

N. Garnett: "Just-in-time: The philosophy of working properly", in *Financial Times,* 20 September 1989, p. 20.

R.J. Graham: *Project management: Combining technical and behavioural approaches for effective implementation* (New York, Van Nostrand Reinhold, 1985).

T. Gunton: *Inside information technology: A practical guide to management issues* (New York, Prentice Hall, 1990).

R. Hinton: *Information technology and how to use it: A handbook of effective practice* (Cambridge, ICSA Publishing, 1988).

R. Hirschheim: *Office automation: A social and organizational perspective* (New York, John Wiley, 1985).

A. Hodgson: "Deming's never-ending road to quality", in *Personnel Management*, July 1987.

A.A. Huczynski and D.A. Buchanan: *Organizational behaviour: An introductory text*, 2nd edition (Hemel Hempstead, United Kingdom, Prentice Hall, 1991), p. 357.

Ingersoll Engineers Ltd.: *Technology in manufacturing* (Rugby, United Kingdom, 1987).

International Organization for Standardization: *ISO 9000: Quality management*, 5th edition (Geneva, ISO, 1994).

G. Kanawaty: *Introduction to work study* (Geneva, ILO, fourth revised edition, 1992).

A.T. Kearney: *The barriers and opportunities from information technology: A management perspective* (London, 1984).

J.P. Kotter and L.A. Schlesinger: "Choosing strategies for change", in *Harvard Business Review*, Vol. 57, No. 2, 1979, pp. 106-114.

R. Long: *New office information technology: Human and managerial implications* (London, Croom Helm, 1987).

I. McLoughlin and J.H. Clark: *Technological change at work* (Milton Keynes, United Kingdom, Open University Press, 1989).

E. Mumford: "Participative system design: Structure and method", in *Systems, Objectives, Solutions*, Vol. 1, No. 1, 1981, pp. 5-19.

—; M. Weir: *Computer systems in work design: The ETHICS method* (London, Associated Business Press, 1979).

D.F. Noble: "Social choice in machine design: The case of automatically controlled machine tools", in A. Zimbalist (ed.): *Case studies on the labour process* (New York, Monthly Review Press, 1979), p. 44.

N. Oliver: "The dynamics of just-in-time", in *New Technology, Work and Employment*, Vol. 6, No. 1, 1991, pp. 19-27.

A.M. Pettigrew (ed.): *The management of strategic change* (Oxford, Basil Blackwell, 1988).

—: "Introduction: Researching strategic change", in A.M. Pettigrew (ed.): *The management of strategic change* (Oxford, Basil Blackwell, 1988), pp. 1-13.

D.A. Preece: *Managing the adoption of new technology* (London, Routledge, 1989).

United Kingdom. Department of Trade and Industry: *The DTI pilots: Evaluation results* (London, HMSO, 1985).

R.W. Walton and G.E. Susman: "People policies for the new machines", in *Harvard Business Review*, No. 2, March-April 1987, pp. 98-106.

R.H. Waterman: *The renewal factor* (London, Transworld, 1987).

THE NEXT STEPS   8

Nigel Corlett

As we said in the introduction, an aim of this book is was to introduce to people working in business and industry a direction for improving their effectiveness. This embraces an integration of technical advances and knowledge about people so that the people can give of their best to exploit the technology, rather than attempting to do it the other way round, as so often happens.

To do this, we have brought together proven knowledge from a number of fields of science and shown how they may be used in an integrated way. It is not a question of one thing (e.g. just technology) or another, but a recognition that, when one is introduced, it opens up possibilities for improvement via some of the other areas. But the overriding message is that people are our major resource. With their abilities to plan, to decide, to understand and to learn, and with well-recognized links between work conditions and the maintenance of interest and motivation, they are the foundation on which any organization can grow, providing it recognizes these facts and implements the ideas in relation to the size of the organization, its products and the conditions of the country in which it is based. Some limitations due to knowledge will also obviously exist. But these are surmountable. Many countries have an ergonomics society, consisting of people who are interested in the subject as well as those who are competent in it. Colleges and universities are introducing the subject, very often in departments of industrial or manufacturing engineering or in departments of psychology, or in departments of operations or human resource management. Also there are groups in psychology or sociology interested particularly in problems of work organization who can be called upon for help.

Of course there are always books and journals, although some are more readable than others. Reference in the various chapters should help you in identifying the relevant ones for your specific area of interest.

There is an increasing number of consulting organizations offering advice in this field. In the context of this manual, we would urge that outside expertise is seen as a resource for information, training, stimulation or guidance, and not as running the changes and a substitute for the company's own personnel. Imposed solutions do not stick, nor do people learn from them how to carry on the development for themselves.

Whichever route you choose to improve and develop, recall that the message throughout this manual is that people are our key resource. With knowledge, training and motivation people can climb Everest, walk across the Arctic, sail the Atlantic or round the world. Design situations so that people can concentrate on doing the job, rather than overcoming obstructions to working, and design them with the experts, who are the people who will have to work there. If people are clever enough to run their homes, look after their families and pursue their hobbies, then they are clever enough to accept a lot more responsibility and authority to decide about and run their own work. Being treated like adults is necessary for adults; being treated like children or, even worse, human machines of little consequence, will create frustration, opposition and covert obstructive behaviour.

The experience of work is not just a personal or localized phenomenon, of interest only to the individual and perhaps his or her colleagues and immediate manager. As was noted earlier in this book, organizational structures inherit a lot from the culture in which they were developed. What people want out of work is also influenced by the culture of the society in which they live, and that in turn affects how they will consider changes in their experience of work.

So when we talk about cooperative working, of creating jobs which produce recognizable, testable units which evidently contribute to the whole product, and urge the arrangement of authority so that people can take any action necessary to keep on doing their job well, it is not just goodwill or sentiment. It is a clear recognition, based on both research and application, that as people come to work bearing all the influences of their social life and background, as well as their own human needs and wishes, we should structure the work in their – and our – lives to match in many ways the people we are.

That is why this publication has not just explained technology or payment systems or production control. It has put together a number of key, interacting factors which are relevant to people's ability to perform well. No one factor is more important than another, although, in different circumstances, different factors will predominate. They all influence each other, and each affects the decision which can be made regarding the others.

Of course, this book will not answer everyone's problems; as was said earlier, complex problems require complex answers. There is much more to know and to apply, but much is already clear, from the many studies and projects from around the world. What we hope is clear from this manual is that a dedicated determination to improve, with the involvement of those who have to work with the improvements and the back-up of specialists where necessary, will in fact create major benefits which will not be just a one-off change, but a continuation of improvements over many years.

Figure 8.1 provides a summary of the content of this book and illustrates how the various chapters integrate to provide a coherent system for improving work. It will now, we hope, be clear that work must be possible to do, but also that each individual must be motivated to do it. If we keep in mind that it is people, not machines, which do the work, and arrange things so that people can keep the technology doing what it is supposed to be doing, then we will be well on the way to a successful and long-term future for our organization.

Figure 8.1   Designing for quality work

Making the work possible to do

Matching the physical and mental demands of the work, and of the physical environment, to people's capacities and abilities

(Chapters 2, 3, 4 and 6)

Matching the technology, supply and operating systems to the circumstances of the company and its markets

(Chapters 4, 6 and 7)

Making the work worth doing

Arranging the content of jobs so that people can be responsible for a recognized unit of work, and have authority over how it is done and the circumstances under which it is done (e.g. basic maintenance, material supply)

(Chapters 2, 3, 4 and 5)

Arranging the work organization and the environment so that they support the needs of those doing the work, provide recognition, fulfil training and advancement needs, and ensure cooperative decision-making among those involved in a problem

(Chapters 5 and 7).

A more possible job

A more motivational job

A better quality of job, work and output